Environmental Division 2017

AA009264

Core Programming Area at the 2017 AIChE Spring Meeting and 13th Global Congress on Process Safety

San Antonio, Texas, USA
26 – 30 March 2017

ISBN: 978-1-5108-4144-4

Printed from e-media with permission by:

Curran Associates, Inc.
57 Morehouse Lane
Red Hook, NY 12571

Printed by Curran Associates, Inc. (2017)

For permission requests, please contact AIChE
at the address below.

AIChE
120 Wall Street, FL 23
New York, NY 10005-4020

Phone: (800) 242-4363
Fax: (203) 775-5177

www.aiche.org

Additional copies of this publication are available from:

Curran Associates, Inc.
57 Morehouse Lane
Red Hook, NY 12571 USA
Phone: 845-758-0400
Fax: 845-758-2633
Email: curran@proceedings.com
Web: www.proceedings.com

TABLE OF CONTENTS

482030 AGILE Award Keynote Address

Abstract:

Factors in Selecting AGILE Honoree

1. Overall sustained excellent industry leadership in businesses that employ chemical engineers

2. Recognized for efforts in innovation and creating strategies that ensure success

3. Value of the keynote to the meeting attendees and the members of the Institute Schedule

Event	Time	Location
Continental Breakfast	7:00AM - 8:00AM	Henry Gonzalez Convention Center, Lila Cockrell Theatre
AGILE Keynote Address	8:00AM - 9:15AM	Henry Gonzalez Convention Center, Lila Cockrell Theatre

The 2017 AIChE Spring Meeting and 13th GCPS AGILE Keynote Address will be presented by AGILE award recipient John Y. Televantos, Partner, Arsenal Capital Partners.

John Televantos is a Partner at Arsenal Capital Partners, a middle market private equity firm focused on the specialty industrial, healthcare and specialty financial service sectors. Before joining Arsenal in 2006, Dr. Televantos was President of Aqualon Company, owned by Hercules, Inc.; President and Founder of Helios Chemical Company; and Chief Executive Officer for Foamex International.

Dr. Televantos began his career with Union Carbide in 1977 in silicones research. There he developed two key processes for the manufacture of Methyl Chloride and Siloxane polymers that were commercialized. These processes not only improved efficiency and reduced cost but also dramatically reduced waste. He moved through a variety of management positions there and later at ARCO Chemical, where he became vice president of strategy and development.

At ARCO John co-invented and led a team that developed and commercialized a breakthrough technology in polyol manufacture (a key raw material for polyurethanes) eliminating a whole process step along with waste and emissions, while at the same time opening the door to the creation of new higher-performance polyurethanes. The technology has since been licensed by most of the major global polyol manufacturers with significant waste reduction worldwide.

At Arsenal Capital, Dr. Televantos serves as co-head of the specialty industrial practice. During his tenure, he has been part of several acquisitions and successful exits in the chemicals and materials sectors. In addition to serving on the boards of several Arsenal-owned companies, he is lead director on the board of Balchem Corp.

480979 Parameters for Properly Designed and Operated Ground Flares: 2017

Abstract:

Multi-point Ground Flares (MPGF) are the preferred technology for pressure relief in many ethylene plants. Like elevated flares they provide a safe means of depressurizing the plant should an upset occur. Unlike elevated flares proper design criteria have not been widely disseminated or discussed.

Based on a series of performance studies conducted at the UOP Callidus multi-point flare test facility, engineering calculations, and computational fluid dynamics (CFD) models greater insight into the design tradespace and the resulting safe and environmentally sound operation can be gained.

Various design inputs for these flares results are discussed through physical test data and calculation including burner head design, plot space requirements, fence height requirements, and number of stages required for correct operation. The resulting impact on outputs, capital equipment cost, operating cost, environmental impact, and safety are also highlighted.

NOISE TESTING AND PREDICTION METHODS FOR MULTI-POINT GROUND FLARES

By

Scot Smith
Zeeco, Inc.
Broken Arrow, OK, U.S.A.

Brian Duck
Zeeco, Inc.
Broken Arrow, OK, U.S.A.

Justin Roberts
Zeeco, Inc.
Broken Arrow, OK, U.S.A.

Spencer Randolph
Zeeco, Inc.
Broken Arrow, OK, U.S.A.

Reginald Keith
Hoover & Keith
Houston, TX, U.S.A.

Garrett Porter
Hoover & Keith
Houston, TX, U.S.A.

Mr. Roberts joined Zeeco in 2010. As a graduate of Oklahoma State University, Mr. Roberts received a Bachelor of Science in Mechanical Engineering as well as a Bachelor of Science in Aerospace Engineering from Oklahoma State University. He currently serves as a Flare Applications Engineer.

Abstract

As demand for Multi-Point Ground Flares increases both domestically and internationally, noise generation from these flares has become a focal point of discussion. Currently, there are no set industry standards for noise predictions and theoretical noise values, which often vary between flare suppliers.

Over the years, Zeeco has conducted numerous noise measurement tests on Multi-Point Ground Flares. This paper will cover the testing and test results obtained in conjunction with one of the foremost noise consultants in the industry. In addition, this paper will address in detail noise generation from various gases at multiple sonic and sub-sonic flow rates. Further information and analysis will be provided to discuss the impact of jet noise vs. combustion noise, as well as calculating noise at various distances.

Zeeco hopes to set standards within the combustion industry in regards to predicting noise for Multi-Point Ground Flares based on the relationships correlated from the test data.

Presented at the

2017 Spring Meeting and 13th Global Congress on Process Safety
San Antonio, Texas – March 26-30, 2017

Overview of Multi-Point Ground Flares

Developed in the 1970s, Multi-Point Ground Flares (MPGF) derive their name from their physical layout. Instead of locating the flare flame on an elevated structure, the flame is spread out in a grade-mounted field of multiple pressure assisted flare tips. The tips are then arranged in stages that open as the upstream pressure and gas flow increases and close as pressure and flow decreases.

MPGF are often selected for heavy hydrocarbon service with high available pressure; however, they can be used for a wide range of gas compositions. High pressure assists in obtaining full smokeless operation, which can be difficult to do with other assist mediums. Each tip in a MPGF has unobstructed air access, allowing the momentum from the high exit velocity of the flare gas to entrain the necessary air for full combustion. MPGF are designed to provide maximum smokeless performance, while minimizing radiation impacts and the need for a large sterile area around the flare. Installing a fence around the field can block the visibility of the flare flame, serving a dual purpose of reducing radiation and the likelihood that flaring operations will be a nuisance to the public. Figure 1 is an example of a typical MPGF installation.

Figure 1: Typical MPGF Installation

Unlike elevated flares, MPGF provide easy access for maintenance. Typically, all staging equipment is located at grade and outside the fence. As a result, workers can safely perform normal operations near the flare without being affected by a flaring event. Figure 2 illustrates a typical fence installed with a MPGF.

Figure 2: Typical MPGF Fence

Noise

Noise can be defined as "excessive or unwanted sound. In general, any sound that is annoying, interferes with speech, damages the hearing, or reduces concentration or work efficiency may be considered noise. In air, sound is radiated spherically from its source as a compressional wave, being partly reflected, absorbed or transmitted by hitting an obstacle. Noise is usually a non-periodic sound wave, as opposed to a periodic pure musical tone or sine –wave combination. It is characterized by its intensity (measured in decibels), frequency, and spatial variation." (Darling, 2016)

A decibel (dB) is a log base scale developed to quantify sound. There are two common uses of decibel levels. One is sound power (PWL) and the other is sound pressure (SPL).

Sound power (PWL) or acoustic power is the rate at which sound energy is emitted, reflected, transmitted or received, per unit time. Sound pressure or acoustic pressure is the local pressure deviation from the ambient atmospheric pressure, caused by a sound wave. The sound pressure scale usually ranges from 0 to 140 dB. The 0 value of the scale occurs when sound pressure equals the threshold of human hearing. Figure 3 shows a comparison of sound pressure levels and sound pressures for a variety of different common objects.

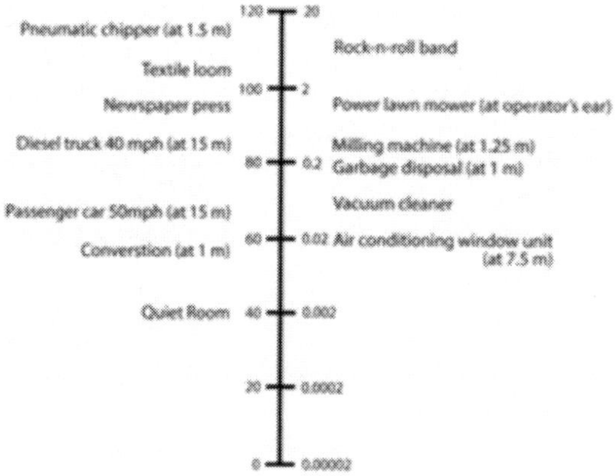

(Breysse, 2006)

FIGURE 3: SOUND PRESSURE LEVEL (SPL) AND SOUND PRESSURE (PA) OF COMMON ITEMS

Even though a decibel has a unit of dB, noise valves are often given in dBA. The "A" designation means the noise level has been modified (i.e. A-Weighted) to de-emphasize the low and very high frequencies which pose less of a risk to hearing. In this paper, all noise values will be shown as unweighted unless stated otherwise.

Another important item when looking at noise data is Leq. When a noise varies over time, the Leq is the equivalent continuous sound which would contain the same sound energy as the time varying sound. In essence, this is the average measurement over a duration of time.

Test Setup

Testing was conducted at Zeeco's test facility in Broken Arrow, Oklahoma. One MPGF flare tip was installed on the test stand which can be seen in Figure 4.

FIGURE 4: MPGF TEST SETUP

With this test setup, gas flow was measured using a 4-inch orifice run. The upstream pressure in the orifice run was measured using a digital pressure transmitter and the temperature was measured using a thermocouple. The differential pressure across the orifice was measured using a differential pressure transmitter.

Tip pressure and gas temperature were also recorded for secondary flow measurement verification. Tip pressure was recorded using a digital pressure transmitter. Gas temperature was recorded using a thermocouple. All aforementioned data was recorded simultaneously using a data acquisition system (DAQ). A weather station was also connected to the DAQ that measured wind speed, wind direction, ambient temperature, barometric pressure, and relative humidity throughout the entire test, which allowed for accurate accounting of atmospheric attenuation in the analysis.

The test fuels used were Tulsa Natural Gas (TNG) and Propane.

Noise measurements were recorded at distances of 100′-0″ and 200′-0″ to the East of the flare tip using two Norsonics NOR140 Type I noise meters. One meter was placed at each distance to measure simultaneously during the test points. Each measurement point lasted 60 seconds.

In order to minimize the amount of background noise, testing was conducted at night with all non-essential equipment (compressors, forklifts, etc.) shut off to avoid contamination of the noise results.

To aide in the testing, Zeeco partnered with Hoover & Keith, a well renowned noise consultant located in Houston, Texas. A consultant from Hoover & Keith was present during all testing and was involved in data analysis.

Results

Background Noise

To ensure accuracy of the test data, background noise points were taken before and after testing. The average ambient sound pressure level was approximately 64 dB. As observed in Figure 5, ambient noise was dominated by low frequencies. While every action was taken to reduce ambient noise, proximity to city streets and highways were uncontrollable factors that likely led to the slightly elevated levels of low frequency sound.

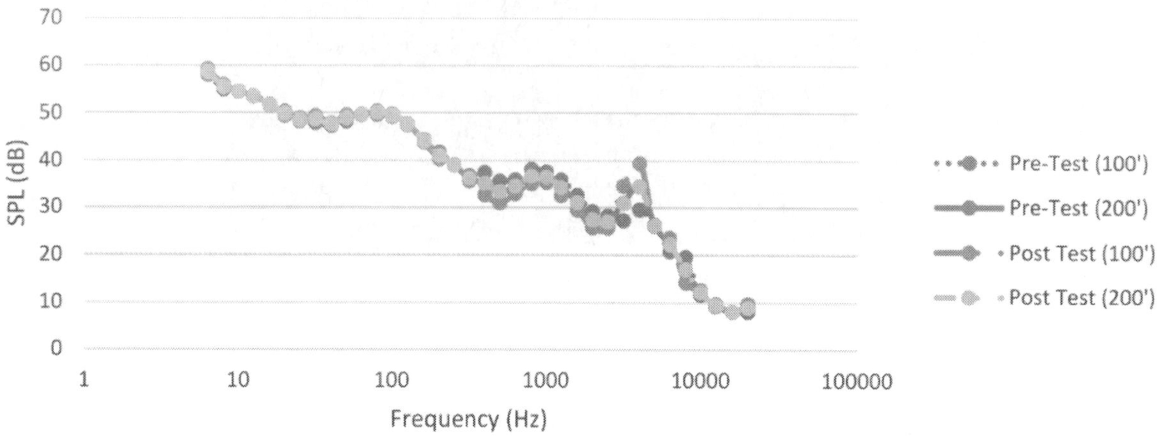

FIGURE 5: AMBIENT NOISE AS SOUND PRESSURE LEVEL IN 1/3 OCTAVE BANDS

Sound Power Levels as a Function of Sonic and Subsonic Flows

All resulting noise data was analyzed on an unweighted basis as sound pressure levels in 1/3 octave bands. All noise measurements were recorded over a 60 second measurement duration. During the measurement duration, a 1/3 octave band spectrum was recorded at 1-second intervals. The 60 measured 1-second 1/3 octave band spectra were then used to compute a single 60-second Leq 1/3 octave band spectrum. The data presented in this paper are the computed 60-second Leq 1/3 octave band spectra or the overall (or total) level derived from these 1/3 octave band spectra.

Shown in Figure 6 is the sound pressure level (SPL) in dB versus fuel flow rate in pounds mass per hour.

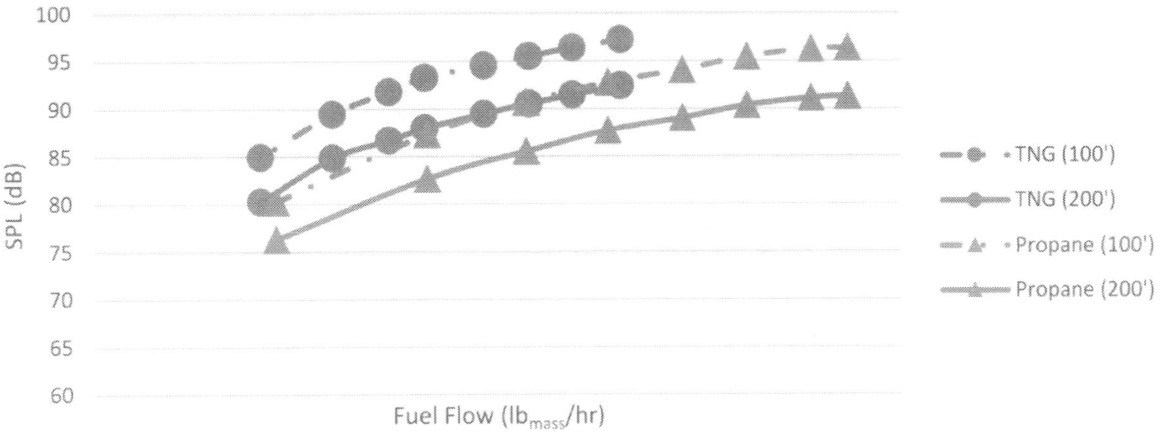

FIGURE 6: SOUND PRESSURE LEVEL (SPL) AS A FUNCTION OF FUEL MASS FLOW RATE

The sound pressure level for each test point was converted to a sound power level using Eq. (1).

$$PWL = SPL + 20Log(r) + 0.5 \quad \text{Eq. (1) (Cunha-Leite, 1988)}$$

Variable "r" is the direct distance in feet from the noise source to the noise measurement location. When converting to sound power level, atmospheric attenuation was taken into account using onsite meteorological data. Upon analysis of the calculated sound power levels, as shown in Table 1 and Table 2, the largest percent difference between the two measurement points was 0.6%, with an average percent difference of 0.4% for TNG. A maximum percent difference of 1.5% was observed for propane with an average percent difference of 0.4%. Each type I meter has an accuracy range of ± 1 dB. When comparing both meters this yields a ± 2 dB range. Therefore, the percent differences in the measurements between the two noise meters fall within the combined accuracy range of the meters. The following results are analyzed as a sound power level. Figure 7 shows a graph of the data from Tables 1 and 2.

TABLE 1: COMPUTED SOUND POWER LEVEL (PWL) – TNG

TP#	PWL - 60 Second Log Average			
	100'	200'	Δ dB	% Diff
1	138.7	139.3	0.6	0.4
2	137.9	138.4	0.5	0.4
3	137.0	137.4	0.5	0.3
4	136.0	136.4	0.4	0.3
5	134.7	134.9	0.1	0.1
6	133.2	133.5	0.3	0.2
7	130.9	131.7	0.8	0.6
8	126.3	127.1	0.8	0.6
Average % Difference			0.5	0.4

TABLE 2: COMPUTED SOUND POWER LEVEL (PWL) - PROPANE

TP#	PWL - 60 Second Log Average			
	100'	200'	Δ dB	% Diff
9	137.8	138.2	0.4	0.3
10	137.7	138.0	0.3	0.2
11	136.9	137.2	0.3	0.2
12	135.5	135.9	0.4	0.3
13	134.2	134.6	0.4	0.3
14	132.1	132.4	0.3	0.2
15	128.7	129.5	0.8	0.6
16	121.7	123.1	1.5	1.2
Average % Difference			0.5	0.4

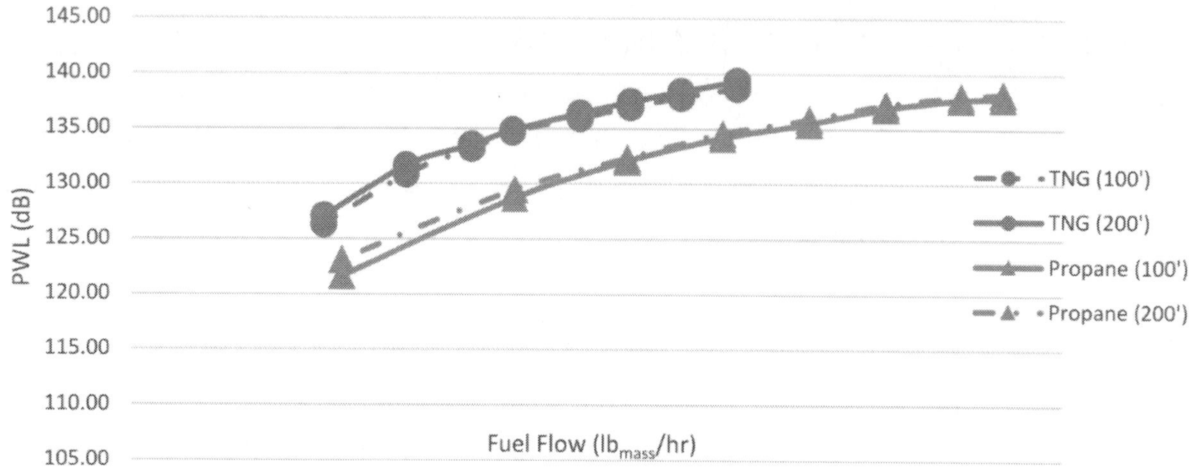

FIGURE 7: SOUND POWER LEVEL (PWL) AS A FUNCTION OF FUEL MASS FLOW RATE

Shown in Figure 8, the sound power levels versus tip static pressure. Also noted are the calculated critical pressures for both TNG and Propane. The critical pressures were calculated from the specific heat ratio of the fuel gas at their respective flowing temperatures per Eq. (2) and correspond to the point at which the fuel gas reaches sonic velocity.

$$p_c = p_2\left(\frac{k+1}{2}\right)^{\frac{k}{k-1}} \qquad \text{Eq. (2)}$$

$P_c = Critical\ Pressure, psia$

$P_2 = Ambient\ pressure.psia$

$k = Ratio\ of\ Specific\ Heat$ (Reed, 2001)

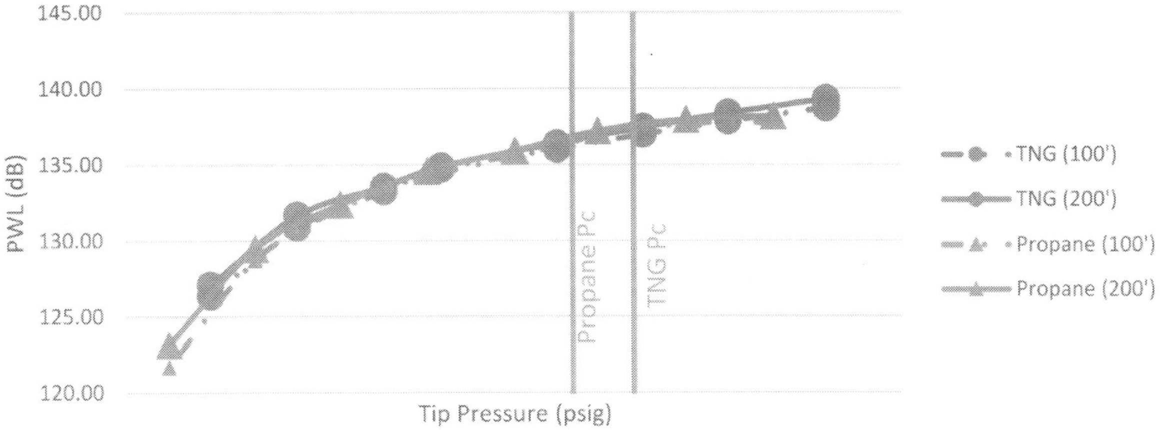

FIGURE 8: SOUND POWER LEVEL (SPL) AS A FUNCTION OF TIP STATIC PRESSURE

Acoustical Efficiency

Using the empirical sound power level, fuel flow rate, and fuel composition, the acoustical efficiency was calculated for each test point. Figure 9 shows the observed trend such that as the fuel flow rate increases, the acoustical efficiency increases for a constant exit area.

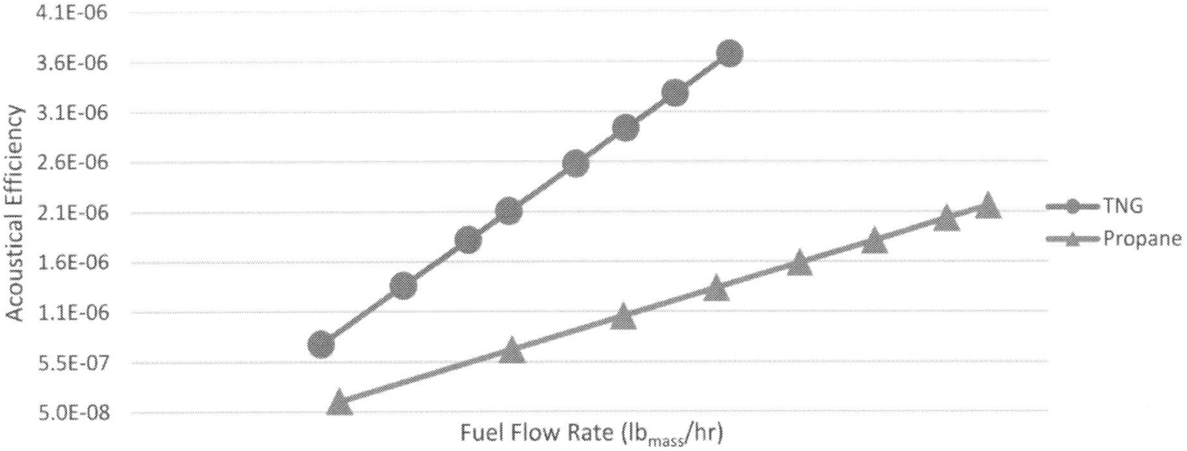

FIGURE 9: ACOUSTICAL EFFICIENCY TREND AS A FUNCTION OF FUEL FLOW RATE

Combustion vs. Venting

Two sets of data points were tested to compare venting jet noise to combustion noise. Shown in Table 3, venting without combustion produced an average 20 dB decrease in sound power level of TNG and an average 23 dB decrease for propane.

TABLE 3: SONIC VS SUBSONIC SOUND POWER LEVEL

Test Point	PWL (Leq dB) at Respective Meter Locations		
	100' (dB)	200' (dB)	Average Δ (dB)
TNG Sonic - Combustion	139	139	20
TNG Sonic - Venting	119	119	
TNG Subsonic - Combustion	136	136	20
TNG Subsonic - Venting	116	116	
Propane Sonic - Combustion	138	138	23
Propane Sonic - Venting	114	117	
Propane Subsonic - Combustion	136	136	23
Propane Subsonic - Venting	112	114	

A further analysis of individual 1/3 octave bands indicates that combustion noise predominantly occurs at frequencies below approximately 500 Hz and jet noise predominantly occurs at frequencies above approximately 2500 Hz. It is important to note that the venting test point low frequency noise could be influenced by the proximity to city streets and highways. The higher trends of the venting case for low frequencies less than 1000 Hz does not appear to correlate to noise mechanisms of the flare, but when compared to background noise frequency appears to be originating from test site surroundings. At these lower frequencies for the venting case, the ambient noise pressure level is higher than that of the flare tip. When analyzing the data as sound power level, the calculation is not applicable to these lower frequencies due to the measured sound pressure level being from ambient surroundings and not the venting point source. When the sound power level is calculated from sound pressure level of ambient noise, the ambient noise is arbitrarily calculated as though it is a point source at 100′ to 200′ from the noise measurement location. This arbitrarily amplifies the sound power level, but is necessary to be applied over the full noise spectrum for the analysis of jet noise in frequencies above 2500 Hz. Figures 10 and 11 show the sound power levels plotted for each 1/3 octave band for both TNG and propane at sonic and subsonic flows.

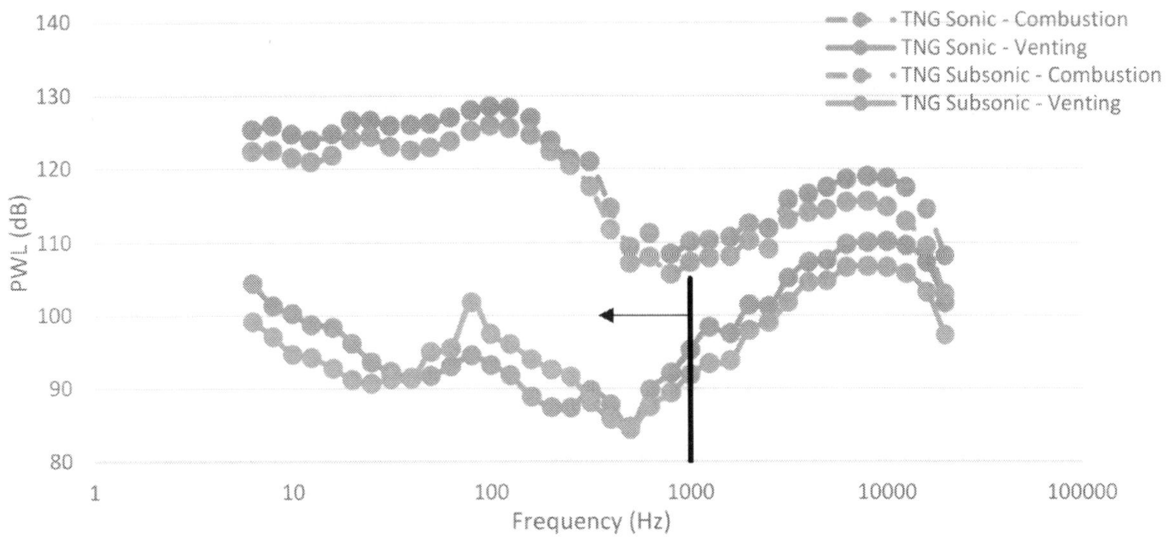

FIGURE 10: SOUND POWER LEVEL (PWL) AS A FUNCTION OF FREQUENCY - TNG

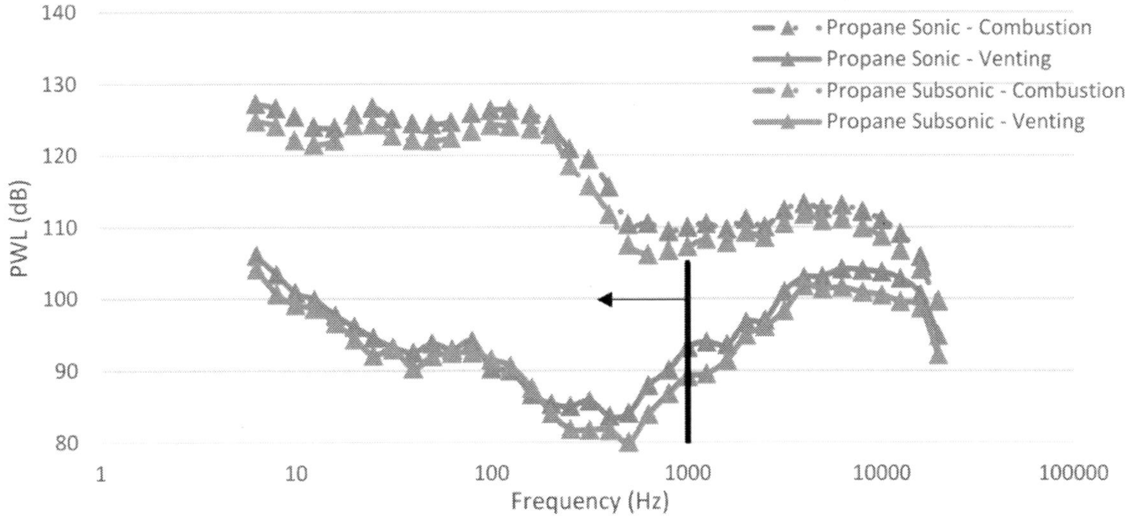

FIGURE 11: SOUND POWER LEVEL (PWL) AS A FUNCTION OF FREQUENCY - PROPANE

In Figure 12 and Figure 13, the sound power level versus 1/3 octave band frequencies have been plotted for each test point, again note the noise spectrum is dominated by below 500 Hz and above 2500 Hz. Individual test point information in these figures are of less concern. The main observable trend is the sound power level increase for low (500 Hz and below) and high (2500 Hz and above) frequencies. This trend is observed in both propane and TNG testing.

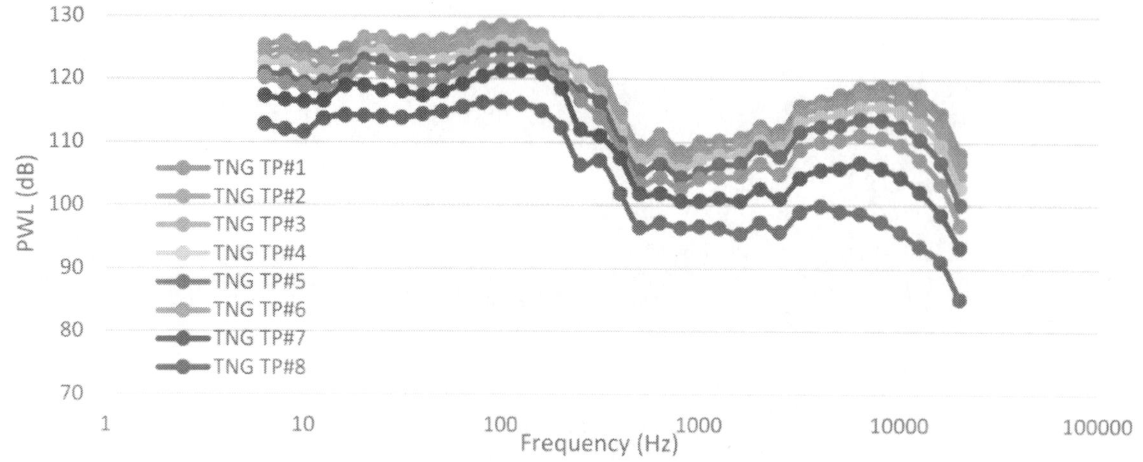

FIGURE 12: SOUND POWER LEVEL (PWL) AS A FUNCTION OF FREQUENCY AND FUEL FLOW RATE – TNG

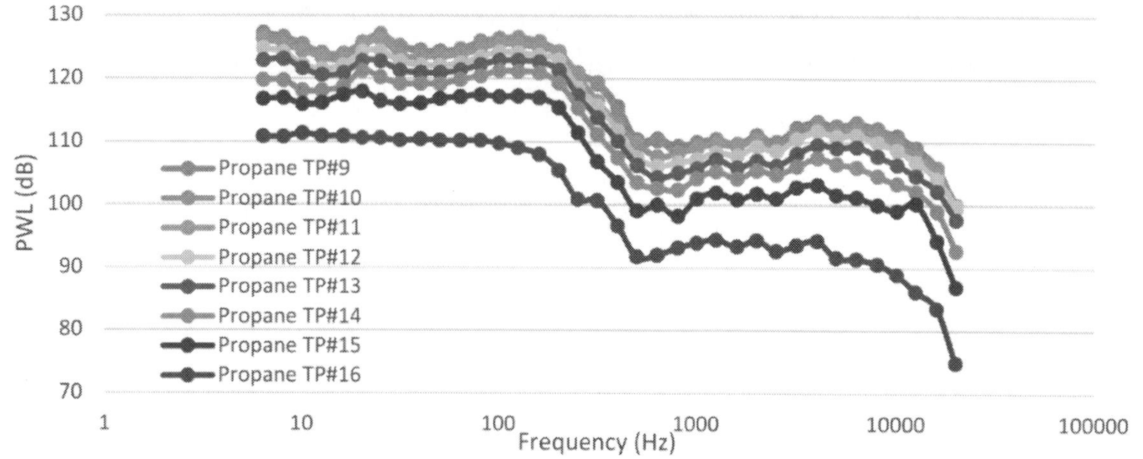

FIGURE 13: SOUND POWER LEVEL (PWL) AS A FUNCTION OF FREQUENCY AND FUEL FLOW RATE - PROPANE

Conclusions

10Log vs 20Log Analysis

Previous information debates using a 10Log versus 20Log relationship to calculate the overall PWL. For each case, a reference fuel mass flow rate and corresponding power level is used to determine a sound power level over a range of fuel mass flow rates. In equations 3 and 4, G_0 references the known fuel flow rate corresponding to a known sound power level. G, is the fuel mass flow rate at which a sound power level is to be determined.

$$PWL_{calculated} = PWL_{reference} + 10Log\left(\frac{G}{Go}\right)$$ EQ. (3) (NARASIMHAN, 1986)

$$PWL_{calculated} = PWL_{reference} + 20Log\left(\frac{G}{Go}\right)$$ EQ. (4) (NARASIMHAN, 1986)

$$PWL_{Calculated} = Sound\ Power\ Level\ at\ Desired\ Flow\ Rate, dB$$

$$PWL_{reference} = Sound\ Power\ Level\ at\ Empirical\ Reference\ Point, dB$$

$$G = Desired\ Fuel\ Flow\ Rate, \frac{lb_{mass}}{hr}$$

$$G_0 = Fuel\ Flow\ Rate\ at\ Empirical\ Reference\ Point, \frac{lb_{mass}}{hr}$$

Figures 14, 15, 16, and 17 show the analysis of the 10Log vs 20Log trends. The actual test data was graphed as well as the data calculated based on G_0 being a low flow point, mid flow point, or a high flow point. These low, mid, and high flow points are all empirical data collected during the testing. The low flow point references the lowest fuel flow point in the data set for that fuel gas, mid flow references a mid-range point, and high flow references the highest fuel flow of the data set for each fuel.

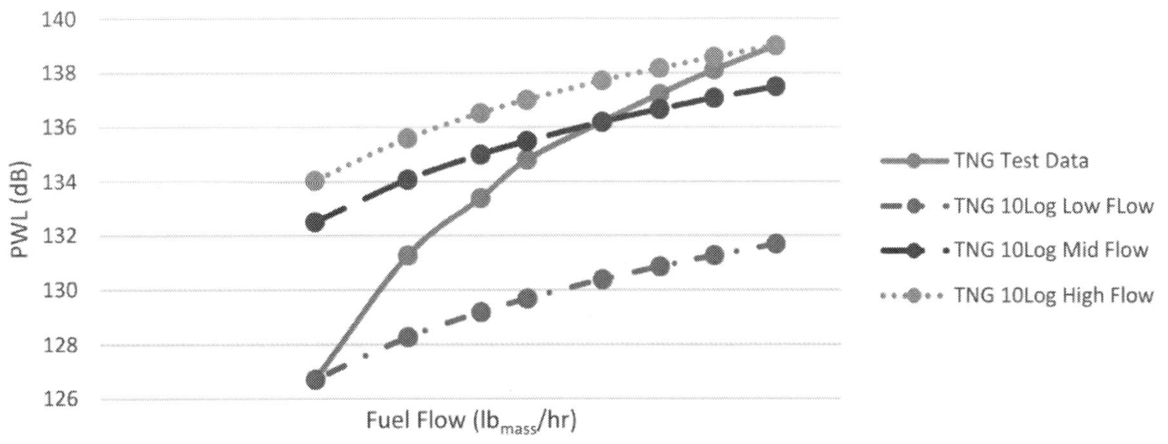

FIGURE 14: 10LOG TREND ANALYSIS - TNG

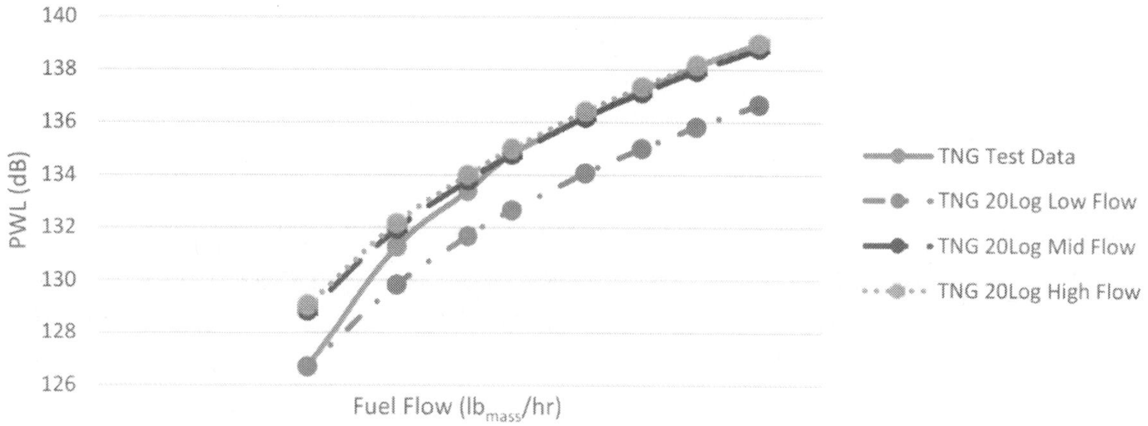

FIGURE 15: 20LOG TREND ANALYSIS – TNG

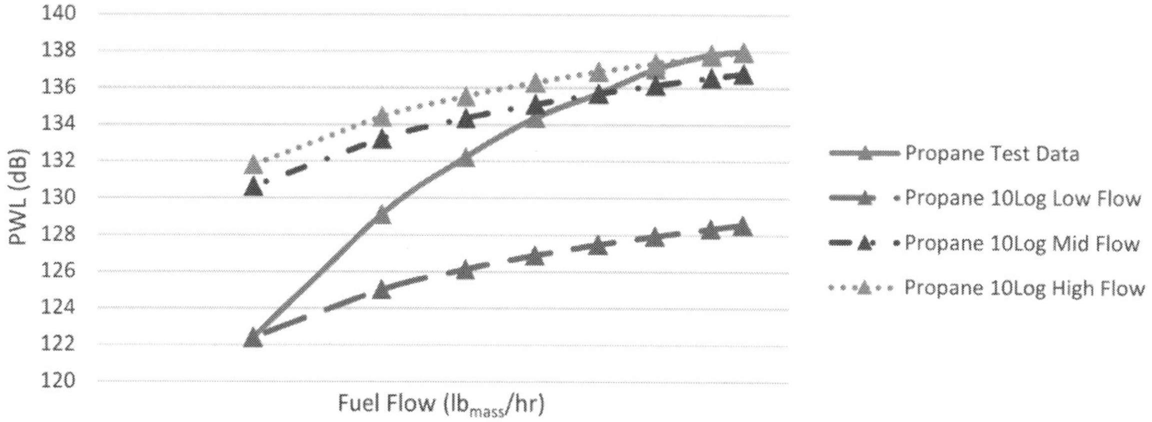

FIGURE 16: 10LOG TREND ANALYSIS – PROPANE

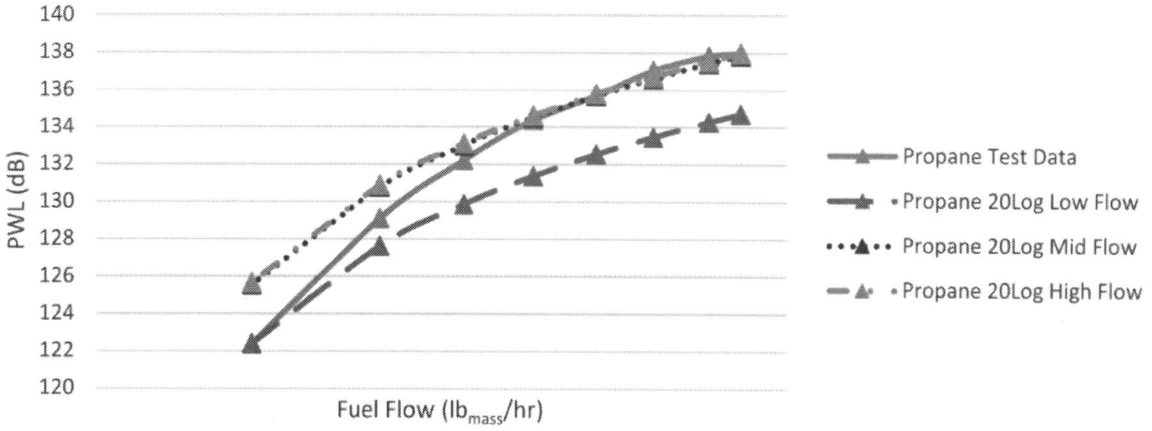

FIGURE 17: 20LOG TREND ANALYSIS - PROPANE

As shown in the figures above, for both propane and TNG, the 20Log trend fit resulted in a smaller percent error. Also observed in the figures above, the referenced empirical data point impacts the percent error. When analyzing a data point corresponding to a smaller mass flow rate, the trend does not accurately represent empirical data.

TABLE 4: PERCENT ERROR IN 10LOG AND 20LOG TRENDS

		Low Flow		Mid Flow		High Flow	
		Avg % Error	Max % Error	Avg % Error	Max % Error	Avg % Error	Max % Error
TNG	10Log	-3.5	-5.3	0.8	4.6	1.9	5.8
	20Log	-1.3	-1.7	0.3	1.7	0.4	1.8
Propane	10Log	-5.0	-6.9	1.2	6.7	2.1	7.7
	20Log	-1.9	-2.6	0.5	2.6	0.6	2.7

As shown in Table 4, the most accurate trend while yielding conservative results, is a 20Log trend based on a data point corresponding to a mid-range mass flow rate. It is important to note that as the calculated sound power level and associated mass flow rate deviate from the referenced point, the percent error increases. This increase is greater at calculated sound power levels associated with low mass flow rates.

Acoustical Efficiency

Multiple publicly available reference articles include discussions on the topic of acoustical efficiencies. Several discrepancies exist between these reference articles as shown in Table 5. The empirical data collected for this testing was investigated for acoustical efficiency trends. Using Eq. (5) and Eq. (6), acoustical efficiencies for each test point were calculated.

$$PWL = 10Log\left(\frac{W}{W_0}\right) \qquad \text{Eq. (5)}$$

$$W = e\left(0.293\frac{hr\cdot W}{Btu}\right) G \cdot H_v \qquad \text{Eq. (6)}$$

$$e = Acoustical\ Efficiency$$

$$W = Sound\ Power, Watts$$

$$W_0 = Reference\ Sound\ Power, Watts$$

$$G = Fuel\ Flowrate, \frac{lb}{hr}$$

$$H_v = Heat\ Value, Btu/lb$$

TABLE 5: ACOUSTICAL EFFICIENCIES FOR A TYPICAL HYDROCARBON

Reference Article	Acoustical Efficiency for a Typical Hydrocarbon
"Predict Flare Noise and Spectrum" (Cunha-Leite, 1988)	$5(10^{-8})$
"Predict Flare Noise" (Narasimhan, 1986)	$1(10^{-6})$
"Noise Generation by Open Turbulent Flames" (Smith, 1963)	$1.23(10^{-8}) - 8.20(10^{-8})$
"Ecological Aspects of Combustion Devices (with Reference to Hydrocarbon Flaring)" (Swithenbank, 1972)	$1(10^{-7}) - 1(10^{-9})$

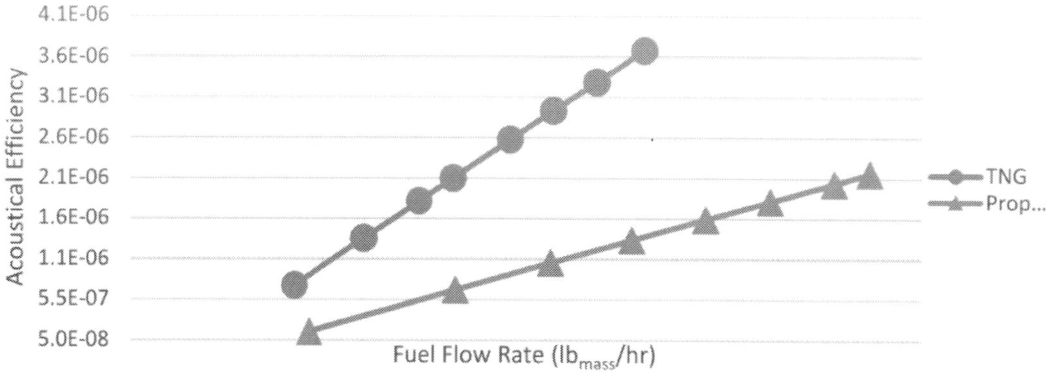

FIGURE 18: ACOUSTICAL EFFICIENCY (E) AS A FUNCTION OF FUEL MASS FLOW RATE

As shown in Figure 18, it was observed that the acoustical efficiency increases with increasing fuel mass flow rate for a constant exit area. Trends of both TNG and Propane were similar with differing rates of change.

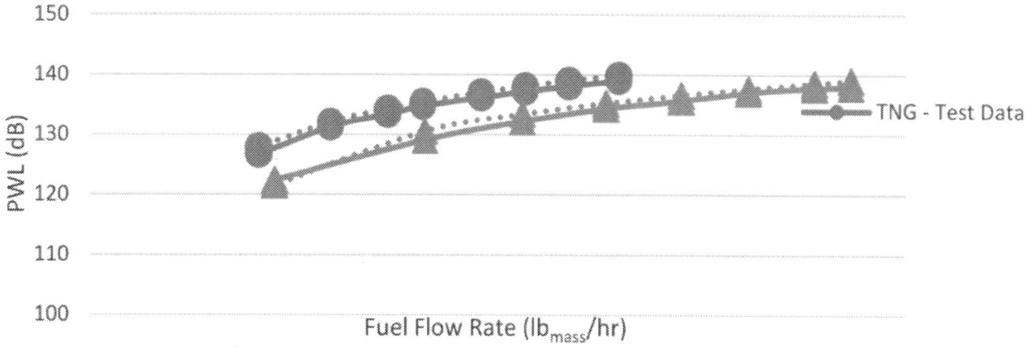

FIGURE 19: SOUND POWER LEVEL (PWL) AS A FUNCTION OF FUEL MASS FLOW RATE

In conclusion, it was observed that as the fuel mass flow rate increased, the acoustical efficiency increased for a constant exit area. When applying the changing acoustical efficiency as a function of fuel composition and flowrate, the prediction model shows a much stronger correlation, as shown in Figure 19. Zeeco is continuing work with Hoover & Keith to determine the impact of these acoustical efficiency trends.

Further Testing

10Log vs 20Log Analysis

While the 10Log vs 20Log analysis shows a more accurate trend correlation when analyzing by means of a 20Log function, testing including higher fuel flow rates would provide a better understanding of the error involved when extrapolating noise values outside of a small range away from the referenced empirical data. A larger range of fuel flow rates would also allow a better understanding of optimal fuel flow rates to use as an empirical reference.

Acoustical Efficiency

With the observance of increasing acoustical efficiencies associated with increasing fuel flow rate for a constant exit area, further testing is required to determine actual causation. Testing of the same format with a multitude of fuel gases would be beneficial. This information would provide more evidence to analyze trends present between fuel gases with different heating values and molecular weights. In addition, fuel blends and inert mixtures would add additional understandings to the phenomena observed. The aforementioned acoustical efficiency testing could potentially yield a more accurate method of predicting multipoint ground flare noise levels.

Combustion vs. Venting

While the combustion versus venting tests provide insight into which frequency ranges are dominated by respective noise mechanisms, it would be beneficial to test a multitude of fuel gases of differing molecular weights and sonic velocities. Adjusting fuel exit areas while maintaining a constant fuel flow rate would yield a better understanding of the driving noise mechanisms of combustion versus venting and would allow a better understanding of the magnitude of impact from combustion noise. This would be facilitated by incrementally decreasing the fuel exit velocity and respective jet noise, while maintaining constant combustion.

References

1. Breysse, P. N. (2006). Noise. John Hopkins School of Public Health.
2. Cunha-Leite, O. (1988, December). Predict flare noise and spectrum. *Hydrocarbon Processing.*
3. Darling, D. (2016). *Noise.* Retrieved from Encyclopedia of Science: http://www.daviddarling.info/encyclopedia/N/noise.html
4. Narasimhan, N. (1986, April). Predict flare noise. *Hydrocarbon Processing.*
5. Reed, R. J. (2001). *North American Combustion Handbook* (Third ed., Vol. I). Cleveland, OH: North American Manufacturing Company.
6. Smith, T. &. (1963, May). Noise Generation by Open Turbulent Flames. *The Journal of the Acoustical Society of America, 35*(5), 715-724.
7. Swithenbank, J. (1972, May). Ecological Aspects of Combustion Devices (with Reference to Hydrocarbon Flaring). *AIChE Journal, 18*(3), 553-560.

479762 Optimization for Flare Operation

Abstract:

Abstract

Flare operation is a challenging task since the flowrate, net heating value, composition of the vent gas and weather conditions vary significantly. The industry is in urgent need of a viable approach to achieve smokeless flaring that can meet the combustion/destruction efficiency standard set by the agencies, while minimize stream and makeup fuel consumption.

In this work, a novel optimization approach was developed based on flare performance models. The controlled variables include flowrate of steam/air assist and flowrate of make-up fuel. For a flaring vent with a specific combustion zone net heating value and particular composition, the most economic solution to achieve smokeless flaring and high combustion/destruction efficiency objective can be identified.

480357 Chemical Kinetics Mechanism Modeling of Sour Gas Combustion

Abstract:

Sour gas combustion is very important for gas processing plants and refineries. Combustion behavior of sour gas is different from other hydrocarbon gases due to the lower heat capacity of sulfur containing components and different radical pool, etc. However, researches on how to combust sour gas stably, effectively, while producing minimal emission is still a big problem. Chemical kinetic modeling provides us an efficient way to assess to this approach. The reaction mechanism is a combination of exciting mechanism of oxidation of C1-C3 hydrocarbons and H_2S. Optimization work is done to compensate the radical pool changes and interaction between C containing and S containing component. The optimized mechanism was validated against real experiment data to achieve a reliable overall mechanism.

480407 Performance Modeling of a Reverse Osmosis Element for Water Purification

Abstract:

Reverse osmosis (RO) membranes provide reliable and efficient separation of dissolved ions from aqueous feeds and are utilized heavily in large-scale water purification operations. Often, these systems will have strict production requirements on pure water as well as complex feed streams containing multiple dissolved ions. To assist with end users' selection process on which RO products best fit their application, FILMTEC™ produces and maintains a software package that can be used to predict the performance of an RO system with regard to operating temperatures, feed pH, ion composition, and production output.* These performance predictions are functions of the various element construction parameters as well as overall system design. In this work, a newly developed membrane element designed for superior antifouling performance is tested over a large concentration range of total dissolved solids. The experimental results are modeled using both empirical and fundamental relationships, and additional performance data is also discussed.

*FILMTEC is a wholly owned subsidiary of The Dow Chemical Company.

SOLID-LIQUID SEPARATION TECHNOLOGY FOR REMOVING CONTAMINANT FINES FROM WATER SCRUBBING, CLARIFIER EFFLUENTS AND GREY WATER

PAPER NUMBER 480801

Barry A. Perlmutter, President & Managing Director
BHS-Sonthofen Inc.
14300 South Lakes Drive
Charlotte, North Carolina 28273-6794
Phone: 704.845.1190
Fax: 704.845.1902
E-mail: barry.perlmutter@bhs-filtration.com
Blog: perlmutterunfiltered.com

ABSTRACT

Treating water for environmental and process benefits is becoming more important in the oil and gas, energy, bioenergy, biochemical and chemical industries. As these operations are looking to become more cost efficient, water usage is being investigated. One critical aspect of these processes is the solid-liquid separation technology that is used to keep the water free of solid contaminant fines.

This paper discusses thin-cake candle filter technology as an alternative to filter presses and bag and filter cartridges for removing solid contaminant fines from the water streams. These contaminants originate from various sources and are generally less than 1 micron (um) in size, which makes their removal very difficult. Candle filter technology and the process of thin-cake building, is a new approach for high-efficiency and cost-effective fines removal.

The paper begins with the problem definition and a discussion of the bench-top laboratory tests that are conducted for problem analysis, process development, technology selection and scale-up. The tests include pressure, filter media, filter aids, and similar process parameters. The candle filter technology is examined for its ability to filter and dry the contaminants to meet the standards of non-hazardous and "no free liquid" for solids disposal. Cake washing is also discussed, as, in some cases, soluble or insoluble components or hazardous solvents must be removed from the solid cake prior to disposal. Three different case histories are presented for purge water treatment units using filter aids, concentration without filter aid, and grey water concentration. Finally, there is a discussion of the impacts of chemical usage in clarifiers on filtration efficiency and troubleshooting.

INTRODUCTION

Water usage has many applications at refineries, gas plants, chemical plants and bioenergy /biochemical facilities. Wet scrubbing at refineries, for example, reduce sulfur dioxide (SO_2/ SO_3) and are the primary and fine particle removal mechanisms from the flue gas. In gasification, water carries the coal fines and catalyst fines during the process. Clarifiers are used in many applications to settle/concentrate to sludge and remove the solids from the process. While the plant location and type of removal process may vary, there is a need to remove the fine particulate matter from the water streams for either reuse or disposal. The particulate matter can be catalyst fines and other contaminants in the feedstock. It is difficult or impossible to remove particles of this size in settling tanks, hydrocyclones or centrifuges, so the particles must be removed by pressure filtration. The use of thin-cake candle filter technology has been proven to be a cost-effective and reliable approach for removing the contaminants, recovering the scrubbing liquids and drying the cake for non-hazardous and "no free liquid" landfill disposal.

PROBLEM DEFINITION

Various catalyst, corrosion products and carbonized particles are carried into the gas and are captured by the water with flow rates as high as 200 m^3/hr. The fine particles are less than 1 micron and can be up to 1% solids in the slurry. Bag and filter cartridges at these solids concentrations are uneconomical and could result in costs of over $500,000 per year. Similarly, the use of filter presses has high operating and maintenance costs. This paper discusses specifically the removal of these fines and provides three illustrative case histories.

TECHNOLOGY OF CLARIFICATION & RECOVERY OF SOLIDS FROM SLURRIES

Candle Filters are installed for clarification and recovery applications from liquids with low solids content. They provide clean fluids to 0.5 micron with either a dried cake (no free liquids) or a concentrated slurry.

Candle Filtration Technology

A candle filter is a pressure vessel filled with tubular filters called filter candles (Figure 1). A typical filter candle is comprised of a dip pipe for the filtrate and pressurized gas, a perforated core with supporting tie rods, and a filter sock (Figure 2). The filtrate dip pipe runs the length of the candle and ensures high liquid flow, as well as maximum distribution of the gas during cake discharge. The tie rods create an annular space between the filter sock and the perforated core, which helps to maintain a low pressure drop during operation and promotes efficient expansion of the filter sock during cake discharge. The filter sock is installed over the candle, and can be made of various synthetic materials, capable of removing particles smaller than 1 μm. As the cake builds during operation, the candle filter's removal efficiency increases, enabling removal of particles as small as approximately 0.5 μm.

Figure 1: Candle Filter Complete

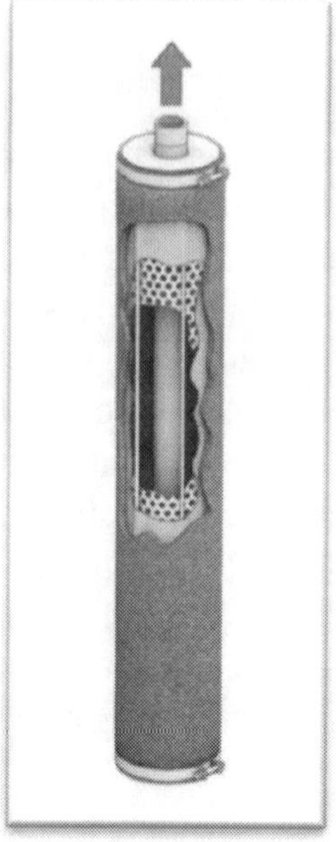

Figure 2: Candle Filter Element

The candles are installed in a pressure vessel constructed of stainless steel or another alloy. Within the vessel are horizontal manifolds called candle registers. Each candle is connected to a register with a positive seal to prevent bypass. Depending on the filter size, each register may contain 1–20 candles and the total candles can be over 200 candles per vessel. The liquid filtrate and pressurized gas flow through the register; automated valves ensure optimum flow in both directions.

During operation, a feed pump forces the slurry into the bottom of the pressure vessel. The solids build up on the outside of the filter sock, while the liquid filtrate flows into the candle, through the registers, and out of the vessel. This process continues until the maximum pressure drop, design cake thickness, minimum flow, or maximum filtration time is reached.

The cake is washed to remove impurities and residual mother liquor, and then dried. Next, low-pressure gas enters the individual candles and expands the filter socks. This process breaks apart the dry cake, which detaches from the filter sock (Figure 3) and falls into the vessel cone. The cake can also be discharged as a concentrated slurry without drying.

Candle filters are used for thin-cake (5–20 mm) pressure filtration applications. They are best suited for filter cakes that are vertically stable because of the orientation of the candles.

Figure 3: Candle Filter Cake Discharge

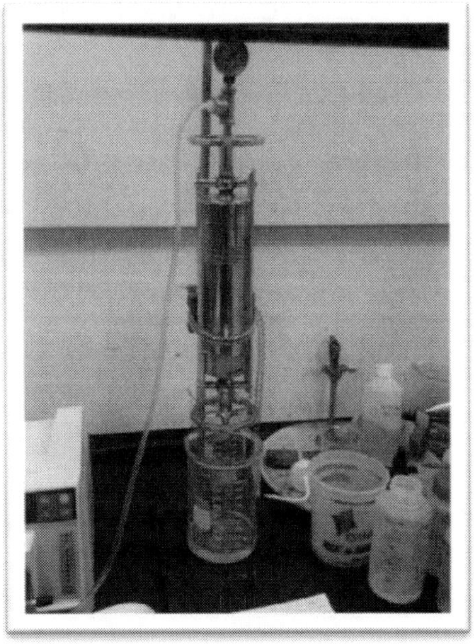

Figure 4: Pocket Leaf Filter Testing

DETERMINING CANDLE FILTER SYSTEM DESIGN

Laboratory testing at a constant slurry flow is used to determine the size and design of a clarification system for processes with extremely low concentrations of solids. The test evaluates the filter media, operating pressure, and cake thickness to determine the optimum clarification system design and size.

A lab-scale leaf filter (Figure 4) is used as the test filter. Various filter media can be installed, depending on the desired filtrate clarity, filtration flux rate (time), cake thickness, and cake discharge rate, as well as compatibility with the process. A peristaltic pump supplies a constant flow of slurry to the filter, and a pressure gage measures the change in pressure (delta P) across the filter media.

The washing analysis consists of running a liquid at the same or different flowrate as the test flowrate through the peristaltic pump, or adding wash liquid to the leaf filter directly and applying air or gas (nitrogen, argon, etc.) pressure until the target pressure drop or quality (conductivity, acid level, etc.) is reached

The cake is then dried with pressurized gas or air. The amount of compressed gas is regulated to prevent the use of excess gas which could lead to overly optimistic moisture targets and uneconomical compressed gas volumes when scaled up.

WATER SCRUBBING AND CLARIFIER EFFLUENT USING FILTER AIDS: ALTERNATIVE TO FILTER CARTRIDGES

In this application, a wet scrubber is used to remove catalyst fines and sulfur dioxide from a fluid catalytic cracking unit flue gas. A purge stream is sent from the wet scrubbing system to a purge water treatment unit and then to disposal. The candle filter system removes catalyst particles fines from the clarifier effluent stream prior to discharge. The flow rate is 70 m^3/hour with a solids concentration of 190 ppm. The particle size is below 0.5 micron down to 0.06 micron. The filtrate specification for discharge is less than 40 ppm. Initially, filter cartridges were planned but the cost of change out alone, based upon the above parameters was estimated to be over several million dollars not including the cost of cartridge inventory, disposal and manpower for cartridge change out, assumed to be every shift.

BHS conducted laboratory tests, as previously described including the use of filter aids due to the size of the particles being less than 0.5 microns. The installation is shown in Figure 5 and consists of two BHS candle filters, each with 100 m^2 of filter area and a filter aid precoat package. The cake discharged is shown in Figure 6 while the filtrate is much below the specification at 5 – 8 ppm.

Figure 5: BHS Candle Filters, 2 x 100 m^2

Figure 6: BHS Candle Filters, 2 x 100 m^2, cake discharge

WATER SCRUBBING-CLARIFIER EFFLUENT CONCENTRATING WITHOUT FILTER AIDS: ALTERNATIVE TO BACKWASHABLE METAL CARTRIDGES

In this application, the wet scrubber water purge is pumped to a clarifier for primary removal of large catalyst solids. The clarifier overflow discharges to an oxidation tower and the effluent from the tower is pumped through the effluent filter and heat exchanger prior to discharge.

The flow rate is 46 m^3/hour with a solids concentration of 200 ppm. The particle size is below 1.0 micron. The filtrate specification for discharge is less than 15 ppm. Initially, backwashable metal cartridges were planned but the low dirt holding capacity of the metal cartridges and the frequent backwash, which is sent back to the clarifier, was estimated to be uneconomical. Therefore, the decision was to look at the BHS candle filter technology solution.

As this process required a concentrated slurry rather than a dried cake, the BHS process solution was concentrating candle filters without filter aid. The installation is shown in Figure 7 and consists of two BHS candle filters, each with 25 m^2 of filter area. The benefits of the candle filter solution included clear filtrates to less than 1 ppm as well as less backwash to the clarifier as the candles are "cleaned" with compressed gas and not liquid.

Figure 7: BHS Candle Filters Downstream of Clarifiers

GREY WATER CONCENTRATING

In this application, grey water containing catalyst fines, coal fines, char and other non-reactive solids required clarification to 0.5 microns. The filtered water is sent back to the gasifier process while the solids are sent to the wastewater plant. The grey water flow rate is 180 m^3/hour with a solids concentration of 200 ppm.

BHS conducted lab testing to determine the filtration flux rate. It was also determined that due to the flow rate and low solids it was a more cost-efficient approach to do the final solids drying in a filter press rather than on the candle filters. Figure 8 shows the final sizing and Figure 9 shows the completed turnkey skid package. The concentrating candle filters produced a clear filtrate of 0.5 microns as well as reduced the feed to the wastewater plant to 10% solids at 300 gallons every 2 hours compared with 800 gpm at 200 ppm.

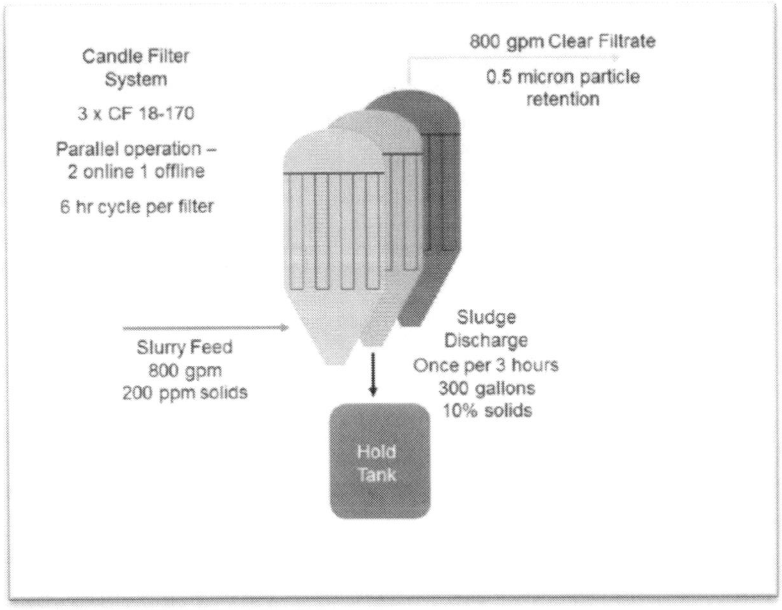

Figure 8: BHS Candle Filter Sizing for Grey Water

Figure 9: BHS Candle Filter Grey Water Installation

FILTRATION SYSTEM TROUBLESHOOTING

There are normally three main areas that must be examined when the filtration system appears not to be functioning properly. These areas include: (1) the filter itself for mechanical reasons, (2) the equipment around the filter is not working and (3) the filter operational procedures are not correct. Let's examine these groups but more importantly, it is necessary to separate the symptoms from the causes. As will be seen, excessive solids in the filtrate (the symptom) has many causes (damaged filter cloth, PSD changes from the reactor, post-precipitation due to temperature changes, etc.).

First, there could be a failure of the equipment itself such an internal components, seals, etc., all as is normally described in the O & M preventative maintenance section from the supplier. This should be the first item to be checked.

Secondly, the filtration system is part of the entire process including the upstream and downstream equipment. For example, are the reactors performing correctly in terms of agitation, temperature control, etc. in order to produce the specified crystals? Are the filter aid systems in tune for mixing, feeding, flow rates, solids loading, etc.? Are the valves and instruments operating correctly and reading the correct variables (calibrations), etc.? Next, what about the pumps that feed the slurry and washing liquids as well as the compressors that feed the gas streams for drying and cake discharge. Finally, are their interlocks in the control system or a control communication problem that are not being recognized that are causing the filter problem.

Finally, there may be process or operational procedures that are resulting in filtration problems. For example, the particle sizes may have changed, the amount of solids in the slurry may have changed, the cake compressibility may have changed, etc. In terms of the operation, has the filtration pressure changed timers changes, speed changed, etc.?

Troubleshooting The Water Scrubbing-Clarifier Effluent Concentrating System

In a clarification application previously discussed, the system was installed and started-up and successfully ran for over one-year. Inexplicitly, the performance changed drastically and the filter media began plugging very quickly during the cycles. The process engineers explained that there had been no change in the process and began to investigate the upstream conditions using the list below.

A. Clarifier overflow with no coagulant / no flocculants
B. Clarifier overflow with only coagulant / no flocculants
C. Clarifier overflow with both coagulant and flocculants
D. Clarifier overflow with only flocculants / no coagulant

After months of work, the process engineers narrowed down the changes to the chemical supplier of the flocculants and coagulants. BHS and the client began lab testing again using the initial slurry along with the chemical supplier and their flocculants and coagulants. The resulting testing showed that the chemical change caused the larger particles to settle out quickly and therefore only the smaller particles reached the filtration system which was blinding the filter media. The client and the chemical company eliminated the flocculants and reduced the amount of coagulant used (better for the client and not so good for the chemical supplier) which produced a smooth and consistent particle size distribution and better filtration rates. Once again, we see the systems and holistic approach to process filtration.

SUMMARY & TAKE-AWAYS

Sherlock Holmes and Dr. John Watson are fictional characters of Sir Arthur Conan Doyle. Process engineers who live in the real-world can learn many things from the two of them for solving process filtration problems. One example that Holmes proves time and again is that there is no benefit to "jumping to conclusions." As discussed in the illustrations, there are many choices of filtration technologies to achieve the necessary level of quality with a cost-effective and a reliable process. Also, in terms of troubleshooting, there are many causes of filtration symptoms. With careful analysis and taking a creative and "non-conventional approach" by looking at what is behind the process data and the operating parameters, the optimum solution will be realized.

480336 Utilization of Flare Gas for Direct Non-Oxidative Production of Liquid Fuels and Chemicals

Abstract:

Natural gas flaring produces 400 million tons of greenhouse gas emissions annually [1]. The current gas-to-liquids technology such as Fischer-Tropsch process requires two stages with synthesis gas ($CO + H_2$) as intermediate, which is extremely capital-intensive and not economically viable for stranded gas utilization. Other technology, e.g., the oxidative coupling process is difficult to scale up. This project will develop a process that enables direct and continuous conversion of flare gas into liquid hydrocarbon fuels. In this process, an initiator was used to facilitate the dissociation of alkanes, which can be converted into liquid hydrocarbons (benzene, toluene, xylenes, mesitylenes, etc) using highly selective zeolite catalysts. This direct conversion will considerably simplify the process, decrease the capital cost and avoid the use of expensive precious metal catalysts and corrosive activators. Therefore, the novel one-step gas-to-liquids process is obviously superior to the current state-of-the-art processes. Moreover, it is expected that the techno-economics of the one-step process will be attractive. This technology has the potential to eliminate flaring for the shale gas industry.

References

[1] http://www.theenergyexchange.co.uk/partner/global-gas-flaring-reduction-ggfr.

480113 Emission Conscious Scheduling of Crude Unloading, Transferring, and Processing for Petroleum Refineries

Abstract:

Emission reduction is becoming increasingly important for petroleum refineries as the increasing concerns of environmental concerns and regulations. Petroleum refinery uses crude oils to produce various fuels such as gasoline, aviation kerosene, diesel, heavy fuel oil, and chemical raw materials such as naphtha and benzene. Crude planning and scheduling is an important part of petroleum refinery due to the potential realization of large cost savings. Cost-effective solution strategies require emission reductions to be addressed from the entire plant point of view, where emission source generations and possible utilizations should be well balanced. But it is a very challenging task due to the large-scale complex manufacturing system, which consists of over 17 major facilities with more than 85 main operation units, and thousands of process streams. The cost-effective emission-reduction strategies should focus on the comprehensively study of material, energy, and information exchanges within the entire manufacturing system, so as to identify the best solutions to the entire plant.

In this paper, a new methodology framework together with a new general scheduling model have been developed for emission conscious crude unloading, transferring, and processing (ECUTP) operations to achieve profitable emission-reduction scheduling (PERS) from the entire system point of view. To identify PERS strategies for refinery plants, a production scheduling model covers both front-end and refinery has to be involved. The key is to optimize the overall material and energy flows to make the plant net profit maximum, meanwhile to ensure emission source generations and utilizations to be smartly balanced. Other manufacturing constraints, such as operation specifications and inventory limits, should also be satisfied as well. Major air emissions from refineries: such as CO_2, volatile organic compounds (VOC), nitrogen oxides (NO_X), and particulate matters (PM) are characterized and quantified.

The scheduling model is a large-scale mixed integer nonlinear programming problem (MINLP). The efficacy of the proposed emission-considered scheduling model has been demonstrated by different case studies. PERS thrusts target on emission source reductions in a profitable and systematic way, which are economically attractive, environmentally benign, and technologically viable for petroleum refineries. It may also help emission generators (not just limited to petroleum refineries) to proactively and systematically reduce their emissions to meet increasingly strict economic and environmental challenges.

480023 Conceptual Design and Cost Estimation of an Offshore Autonomous Once-through Gas-to-Liquid Process Combined with Ammonia Synthesis

Abstract:

Conceptual design and cost estimation of an offshore autonomous once-through gas-to-liquid process combined with ammonia synthesis

A novel process concept is proposed for converting natural gas to liquid Fischer-Tropsch products and ammonia (Figure1). An autothermal reformer with enriched air as oxidant is applied for synthesis gas (syngas) production, and because of the inert nitrogen a once-through Fischer-Tropsch synthesis is the preferred option. In order to maximize the syngas conversion and the production of heavy hydrocarbons, a staged reactor path with distributed hydrogen feed and product withdraw is proposed. The hydrogen is produced by steam methane reforming in a heat exchange reformer (gas heated reformer), heat integrated with the hot effluent stream from the autothermal reformer. The nitrogen needed for ammonia synthesis is produced from air membrane. Tail gas from the last Fischer-Tropsch stage is sent to a gas turbine for power production. The hot exhaust gas from the gas turbine is used for natural gas preheating. The process is autonomous in the sense that it is self-sufficient with power and water, and therefore well suited for production in remote locations such as a floating production unit. The process concept is simple and inexpensive since cryogenic air separation and fired heaters are not required. For the Fisher-Tropsch synthesis microchannel reactors are considered. Cost evaluation is carried out to find out the economic feasibility of the process.

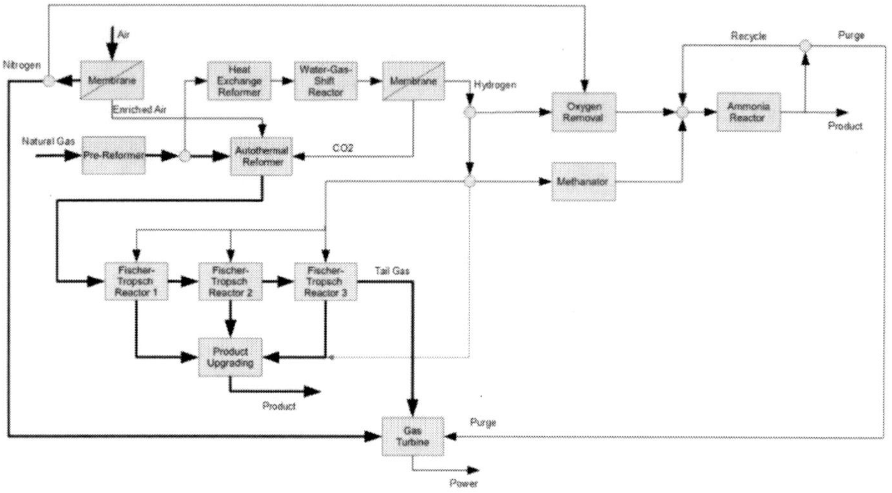

Figure 1: Block flow diagram of the proposed process concept; water and steam are not shown

480897 Study of Liquid Dispersion in Winpak Modular Catalytic Structured Packing: Experiments Vs. CFD Simulations

Abstract:

The response technique was employed to investigate the axial and radial distribution of the liquid phase in Winpak modular catalytic structured packing which was named Winpak-C. Residence time distribution (RTD) experiments were performed at ambient temperature and atmospheric pressure. The axial and radial dispersion coefficients were obtained by solving a two-dimensional dispersion model analytically from the RTD curves. Based on proper assumption, a three-dimensional model of reaction elements in MCSPs was established to determine the dispersion characteristics with the use of computational fluid dynamics (CFD). Porous flow was introduced by using a specification of an extra body force to account for liquid flow through the reaction element. These measured or simulated dispersion coefficients enable the reliable and predictive design and construction of scale-up models.

480396 Exploring the Catalytic Options in Low-Mid Pressure Diesel Hydrotreating

Abstract:

Regularly revised, worldwide diesel fuel specifications continue to press impurity levels lower and product quality higher. Such fuel specification revisions naturally encourage the semi-continuous development of superior hydrotreating catalysts, to satisfy the new requirements and improve the economics of hydrotreacting where applicable. The goal of this talk is to draw rational boundaries where the most active examples of supported NiMo and supported CoMo catalysts ought to be applied, given the inhibiting influences of organic nitrogen species and aromatics species, for example. Included in the scope shall also be application of both types of catalysts simultaneously, where a bed of one of the catalysts is loaded above the other. In consideration of the pressing need for improved hydrotreating catalysts, Honeywell UOP has recently commercialized an entirely new portfolio of supported hydrotreating catalysts under the Unity(TM) trademark. Several of these catalysts shall be featured in this presentation, since they were developed for diesel hydrotreating in the low-to-mid pressure range.

480176 Next-Generation Utilities Optimisation for Refineries and Large-Scale Chemical Production Sites

Abstract:

Refineries and chemical production sites are major consumers of energy in the form of electricity, steam and hydrocarbon feedstocks. Given that tariffs, costs and demands are constantly changing, there is much scope for optimising on-site production, conversion and distribution of energy to minimise cost and emissions, by managing the options available in the most cost effective way while meeting all constraints of the system. However many current 'optimisation' applications do not provide full capabilities for maximising economic value or minimising emissions, particularly in situations where demands from major consumers and the prices of the various components can change on an hourly basis.

This paper describes an advanced optimisation platform for managing and optimising utility operation that not only helps planners rapidly optimise equipment selection and load allocation to improve overall efficiency and reduce emissions and operating costs, but also presents operators with a ranked list of possible actions from which they can choose the best course of action given the situation 'on the ground'.

The approach uses medium-fidelity models of the utilities system and major devices that are coupled with plant operating data via data validation and reconciliation facilities. An advanced optimisation system capable of both continuous and integer decisions determines the economically optimal operating point taking into account equipment and operational constraints, including availability. The resulting mathematical problem is solved within an equation-oriented framework, providing robustness and speed of execution well in advance of most current systems.

Because of the speed of solution, a key feature is the ability to run multiple optimisations and provide operators with a ranked list of potential combinations of changes and their corresponding benefits, within a dashboard tailored for the site. This allows operators clearly to evaluate and discuss which changes are the best to apply when, resulting in advice that is practical and easy to implement and verify. It addresses one of the biggest obstacles to practical realisation of the benefit of optimisation systems, which is gaining operator buy-in to proposed changes.

The paper is illustrated with an examples of implementation on a major European chemical manufacturing site.

478714 Study on Regional Air Quality Impact from Olefin Plant Shutdown Operations

Abstract:

Ground-level ozone is a pervasive air pollutant, which can be potentially aggravated by flaring emissions from olefin plant shutdown operations. Although, flaring is crucial to chemical plant safety, the excessive flaring, for instance the intensive flaring from plant shutdown operations, emits huge amounts of ozone pollution precursors (VOCs and NOx). This will cause tremendous industrial material loss and environmental damage; meanwhile result in potential adverse impact to the regional air quality. Thus, the plant shutdown flare emissions should be minimized at any possibility. In this paper, plant-wide dynamic simulations for flare emissions with regional air-quality modeling are coupled together to quantify the air-quality impact and investigate potential opportunities to mitigate the adverse air-quality impact during plant shutdown operations. It proposed a systematic methodology on air-quality conscious study for shutdown operations of an olefin plant. Through studies of original and optimal cases, it shows that the original plant shutdown operation procedure has a serious and negative effect on the regional air quality; however, the optimal shutdown operation procedure would significantly reduce such impact together with the saving of tremendous emission sources. This study can provide valuable and quantitative supports for decision makings involving multiple stakeholders, including environmental agencies, regional plants, and local communities.

2017 AIChE Spring Meeting & 13th GCPS Global Conference on Process Safety

Up-to-date Engineering Solutions to Estimate and Minimize Thermal Breathing Losses from Storage Tanks

Dr. D. Moncalvo & Dr. M. Davies & Dr. T. Heidermann

Some examples of Tank Vacuum Collapse

Maybe you won't hear about it in the news but it surely costs lots of money!

PROTEGO
for safety and environment

Tank Breathing in API 2000 7th Ed. and ISO 28300

Tank breathing in API 2000 7th Ed. and ISO 28300
is distinguished into Outbreathing and Inbreathing

These standards distinguish between two scenarios:

- Normal in- and out-breathing
  - Liquid pumping
  - Weather changes (sun exposure, rain)
  - Internal failure of heat transfer devices
  - Control valve failures
  - Inert purge and utility failures
  - Vaporization or condensation
  - Steam collapse
  - Other causes
- Emergency out-breathing (fire exposure)

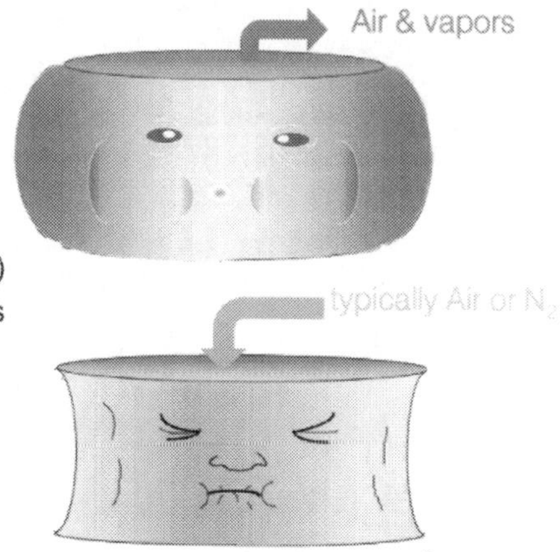

PROTEGO
for safety and environment

Tank Thermal Breathing in API 2000 7th Ed. and ISO 28300

$$\dot{V}_{thermal\ out} = Y \cdot V_{tk}^{0.9} \cdot R_i$$

$$\dot{V}_{thermal\ in} = C \cdot V_{tk}^{0.7} \cdot R_i$$

Insulation factor R_i

$$R_i = 1/[1 + h \cdot l_{in}/\lambda_{in}]$$

Not considering condensation or evaporation of product

Latitude		Y Factor	C Factor			
			Vapor pressure			
			Hexane or similar		Higher than hexane or unknown	
			Average storage temperature			
			< 77°F	≥ 77°F	< 77°F	≥ 77°F
0 - 42°	Houston	0.32	4	6.5	6.5	6.5
42° - 58°	Rotterdam	0.25	3	5	5	5
58° - 90°	Norway	0.20	2.5	4	4	4

PROTEGO
for safety and environment

Tank Thermal Breathing in API 2000 7th Ed. Annex A (*API 2000 5th Edition*)

Tank capacity	Thermal breathing (*SCFH*)	Max. heat transfer rate
Between 20 000 und 180 000 bbl.	2 SCFH each ft² of roof and shall area	Between 50 and 100°F/h
Below 20000 bbl.	1 SCFH each bbl. of tank volume	100°F/h

Annex A method is applicable, if following ***Limitations*** are satisfied:

Tank volume is less than 180 000 bbl.
Max. operating temperature in vapor space is 120°F
Tank is uninsulated
Temperature of stored products and feed is below boiling temperature at max. oper. pressure

Not considering condensation or evaporation of product

PROTEGO
for safety and environment

Tank thermal breathing in the scientific literature

Model	Year	Thermal modeling	Evaporation modeling	Condensation modeling	Sensitivity analysis	Measurements
Sigel et al.	1983	X		X		X
Förster et al. (*PTB model*)	1984	X				
Fullarton et al.	1987	X	X	X		X
Holtkötter et al.	1997	X		X		X
Salatino et al.	1999	X				
Brooks (DIERS pres)	2012				X	
Davies and Moncalvo (1)	2016	X		X	X	
Davies and Moncalvo (2)	2016	X	X	X		X

(1)	Journal of Loss Prevention in the Process Industry , volume 43, doi: 10.1016/j.jlp.2016.06.006
(2)	12th Global Congress on Process Safety (GCPS), Houston, April 2016, ISBN: 978-0-8169-1096-0

PROTEGO
for safety and environment

Davies and Moncalvo (2016)

- Balance of energy within the tank

$$\left(M_{vap}c_{vap} + M_{liq}c_{liq} + M_{air}c_{air}\right)\frac{dT_B}{dt} = -h_{int}A(T_B - T_E) - \dot{M}_{air}c_{air}(T_B - T_{amb})$$

- Balance of energy at the tank walls

$$M_E c_E \frac{dT_E}{dt} = h_{int}A(T_B - T_E) - h_{rain}A(T_E - T_{rain}) + \rho_B \dot{V}_{cond}\Delta h_{cond}$$

- Balance of mass within the tank

$$-\left(V_{air} + V_{vap}\right)/T_B \cdot dT_B/dt + \dot{V}_{cond} = \dot{V}_{air}$$

- Condensation model

$$\dot{V}_{cond} = \beta \cdot 1/p_B \cdot dp_{sat}/dT \cdot [T_B - T_E]A$$

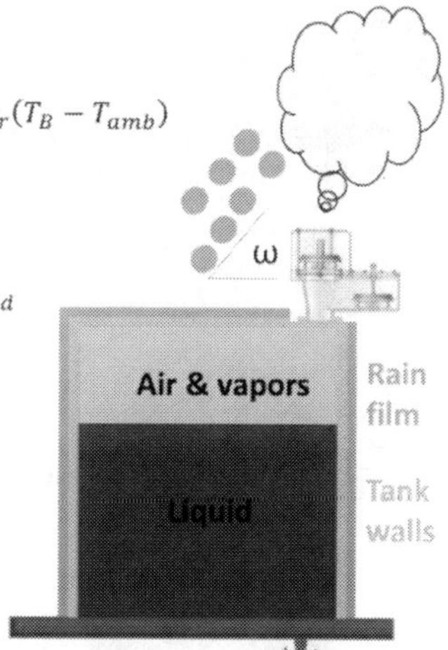

Source	Journal of Loss Prevention in the Process Industry, volume 43, doi: 10.1016/j.jlp.2016.06.006

PROTEGO for safety and environment

Some realistic examples of thermal inbreathing

	Tank 1	Tank 2
Tank height	49.2 ft.	
Tank diameter	49.2 ft.	82.0 ft.
Tank volume (V_{tank})	16,650 bbl.	46,250 bbl.
Initial temperature (T_0)	120°F	
Ambient temperature	60°F	
Average storage temperature	68°F	
Rain intensity	46.1 lb/h ft²	
Rain inclination	68°	
Wall material and thickness	Stainless Steel; $^{1}/_{5}$ in	
Insulation	Mineral wool	
Stored product	Dry Air Saturated Humid Air Saturated Heptane / Air Saturated Methanol / Air	

PROTEGO for safety and environment

Results for ***uninsulated*** *tank 1 (D = 49.2 ft.)* **and** *tank 2 (D = 82.0 ft.)*

Required Inbreathing [SCFH]	API 2000 Main body (max)	API 2000 Main body (min)	API 2000 Annex A	PTB (numerical)	Davies Moncalvo *without* condensation	Davies Moncalvo *with* condensation
Dry air				30,931	31,258	31,258
Humid air	37,179	23,237	16,659	30,730	30,682	48,138
Sat. Heptane / Air				19,576	19,608	31,640
Sat. Ethanol / Air				24,761	24,794	52,825

Required Inbreathing [SCFH]	API 2000 Main body (max)	API 2000 Main body (min)	API 2000 Annex A	PTB (numerical)	Davies Moncalvo *without* condensation	Davies Moncalvo *with* condensation
Dry air				68,807	69,576	69,576
Humid air	76,013	47,508	35,910	67,978	68,149	102,569
Sat. Heptane / Air				41,455	41,680	65,512
Sat. Ethanol / Air				53,485	53,807	108,608

PROTEGO
for safety and environment

Worst realistic cases for rain intensity on *Tank 1*

Required Inbreathing [SCFH]		Rain intensity [lb/h ft²]		API 2000 Main body (max)	Davies Moncalvo *without* condensation
Dry air Tank 1	Scenario 1	46.1	API 2000	37,179	31,258
	Scenario 2	62.4	US historical highest (Holt, MO)		34,020
	Scenario 3	85.9	Worldwide hist. highest (Cherrapunji, IND)		36,910

Lesson learned:

API 2000 7th Ed. Main Body is safe for non-condensing products even under extreme weather

PROTEGO
for safety and environment

Reducing Tank breathing: Insulation Layer and Tank Wall Thickness on *Tank 1*

Required Inbreathing [SCFH]	Insulation Thickness [in]	API 2000 Main body (max)	Davies Moncalvo without condensation
Dry air Tank 1	uninsulated	37,179	31,258
	1	8,905	1,364
	2	5,058	949
	4	2,714	551

Required Inbreathing [SCFH]	Wall Thickness [in]	API 2000 Main body (max)	Davies Moncalvo without condensation
Dry air · Tank 1	0.2	37,179	31,258
	0.25		28,989
	0.3		27,221
	0.4		24,404

Lessons learned:

Increasing tank wall thickness may not suffice to reduce inbreathing requirements

Perform an engineering and economical study to find out the best insulation strategy for your tank

PROTEGO for safety and environment

Tests on the influence of overpressure on tank leakage losses

- Leakage, tested with PROTEGO® portable test rig, may begin at 75% of set pressure as per API 2000 5th Edition Appendix C
- The valves were supposed to be fully open at 1 oz./in² (4.3 mbar) gauge.
- The first test was performed at 100 % overpressure
- The second test was performed at 10% overpressure

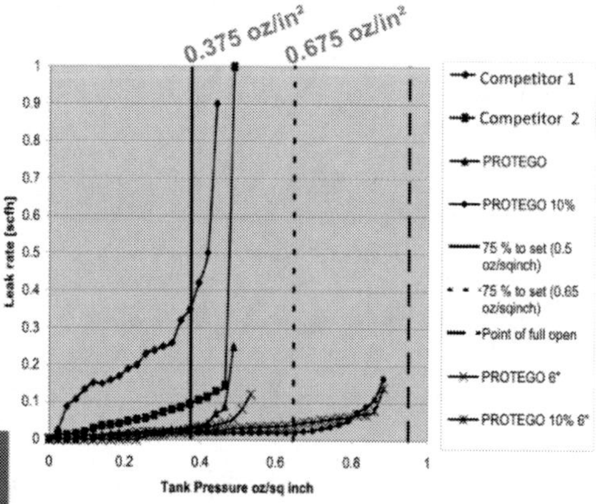

Relieving pressure [oz/in²g]	Overpressure ($\%_{overpr}$)	Set pressure [oz./in² a]	Leakage Pressure [oz./in² g]
1.0	100% (1)	0.5	0.375
1.0	10% (0.1)	0.9	0.675

$$p_{rel\,g} = p_{set\,g}(1 + \%_{overpr})$$

$$p_{leak\,g} = p_{set\,g} \cdot \%_{leak}$$

PROTEGO for safety and environment

Impact of test results in improvement of API 2000 standard

- Two main results from this research test:
  - Avoid choice of valves requiring large overpressures
  - Leakage pressure at 75% set pressure underestimates leakage losses

- Improvement of API 2000 Standard between 5th (1998) and 7th Edition (2014)

Requirements	API 2000 5th Edition	API 2000 7th Edition
Leakage pressure	75 %	90 %
Blowdown	0 %	Dependent on design

- Additional drawbacks to be considered when selecting a PRV with large overpressures:
  - Leakage pressure of PRV may be higher than set pressure of blanketing devices
  - PRV with low set pressures tend to cycle more (additional cost of spare parts)

PROTEGO
for safety and environment

Conclusions

API 2000 7th Ed. Main Body is safe for tanks with non-condensing gases even under extreme weather

Do not underestimate condensation in atmospheric tanks and perform an engineering study, especially if your tank contains volatile products, like (m)ethanol or gasoline

Increasing tank wall thickness may not suffice to reduce inbreathing requirements

Perform an engineering and economical study to find out the best insulation strategy for your tank (trade-off between costs of insulation purchase and maintenance against savings on breathing losses)

An effective minimization of product losses suggests the selection of PRV with minimum overpressure: the authors support the 10% overpressure as standard also for PVRV

PVRV with low overpressures usually guarantee also a chatter-free opening and avoid conflict with blanketing valves and ERV.

PROTEGO
for safety and environment

Thank you for your kind attention!
Reproduction, in part or in full, subject to prior written approval of Braunschweiger Flammenfilter GmbH.
Braunschweiger
Flammenfilter

481850 Environmental Impact of Mining

Abstract:

Mining can have adverse effects on surrounding surface and groundwater if protective measures are not taken. The result can be unnaturally high concentrations of some chemicals, such as arsenic, sulfuric acid, and mercury over a significant area of surface or subsurface.[3] Runoff of mere soil or rock debris -although non-toxic- also devastates the surrounding vegetation. The dumping of the runoff in surface waters or in forests is the worst option here. Submarine tailings disposal is regarded as a better option (if the soil is pumped to a great depth).[4] Mere land storage and refilling of the mine after it has been depleted is even better, if no forests need to be cleared for the storage of the debris. There is potential for massive contamination of the area surrounding mines due to the various chemicals used in the mining process as well as the potentially damaging compounds and metals removed from the ground with the ore. Large amounts of water produced from mine drainage, mine cooling, aqueous extraction and other mining processes increases the potential for these chemicals to contaminate ground and surface water. In well-regulated mines, hydrologists and geologists take careful measurements of water and soil to exclude any type of water contamination that could be caused by the mine's operations. The reducing or eliminating of environmental degradation is enforced in modern American mining by federal and state law, by restricting operators to meet standards for protecting surface and ground water from contamination. This is best done through the use of non-toxic extraction processes as bioleaching. If the project site becomes nonetheless polluted, mitigation techniques such as acid mine drainage(AMD) need to be performed.

The five principal technologies used to monitor and control water flow at mine sites are diversion systems, containment ponds, groundwater pumping systems, subsurface drainage systems, and subsurface barriers. In the case of AMD, contaminated water is generally pumped to a treatment facility that neutralizes the contaminants.[5]

A 2006 review of environmental impact statements found that "water quality predictions made after considering the effects of mitigations largely underestimated actual impacts to groundwater, seeps, and surface water".[6]

THE POTENTIAL USE OF SOLAR PHOTOCATALYTIC OXIDATION FOR REMOVAL OF EMERGING PHARMACEUTICALS FROM EFFLUENT OF WASTEWATER TREATMENT PLANT: A PILOT PLANT STUDY

Fares Almomani, Rahul Bhosale, Anand Kumar, Majeda Khreisheh, Ujjal Gosh
Department of Chemical Engineering, Qatar University, P.O box 2713, Doha, Qatar

Abstract

The treatment of effluent wastewater containing pharmaceutical by solar advanced oxidation processes were studied in a pilot plant. The removal efficiency of four groups of pharmaceuticals (antibiotics, estrogens, acidic, and neutral), commonly detected as emerging contaminants, were tested by different solar-driven photochemical processes including solar -photolysis, solar-photocatalytic oxidation using Fe(III) or TiO_2 and solar based photo-Fenton reaction. Results showed that Solar-photolysis is not effective for the complete removal of the selected contaminants, on other hand solar-photocatalytic oxidation using TiO_2, solar based photo-Fenton reaction showed significant removal efficiencies and high mineralization.
Solar photo-catalytic systems produced a final solution with lower toxicity if compared with original solution.

Keywords: photons, emerging contaminates, solar excitation, hydroxyl radicals

Introduction

Incident presence of what is called emerging contaminates(ECs)such as pharmaceuticals, pesticide in water and effluents of wastewater treatment plants have raised a lot of concern regarding the performance of conventional treatment processes in removing and treating these contaminants (Ternes, 2001, Kolpin et al., 2002). It was reported that the main reason for the presence of these ECs is related to the continual release of these contaminates into this eco-system from different point sources (Huber et al., 2004). ECs are considered non-biodegradable, and consequently it is expected that their concentration in the environment will continue to increase unless they are effectively treated and eliminated. Reasonable concentration of these contaminates have been detected in secondary treatment effluents of wastewater treatment plants (WWTPs) (Kolpin et al., 2002). Contamination of groundwater with ECs was reported from different sources(Ternes et al., 2002). Thus, there is a big need for an effective and cheap post-treatment to remove ECs from wastewater before discharge to the environment.

ECs can be effectively removed from water solution using different treatment processes such as activated carbon, nano-filtration and reverse osmosis(Rosenblatt et al., 1967) (Reddersen et al., 2002, Webb et al., 2003). However, the cost associated with these treatment alternatives limits their applicability.

Advanced oxidation processes (AOPs), on other hand, can be employed as an effective and promising technology for the removal of any traces of ECs from water and wastewater. AOPs which based on the production of highly oxidative radicals are considered as destructive processes. These processes have the capacity to mineralize different contaminates from wastewater in a short period of time. With such technologies, problematic compounds can be partially or totally oxidized into smaller by-products that can be easily removed in conventional biological processes. Applying this strategy would potentially reduce the size of both oxidation and biological reactors, and may reduce both capital and operating costs. This synergy has the potential to open up a new niche in the water treatment field for treating contaminants which are conventionally reserved for other treatment technologies (e.g. activated carbon). The high cost associated with the use of AOPs (reagent and UV-light) limit their applicability. However, most recently, solar advanced oxidation processes were effectively used in the treatment of water and wastewater containing non-biodegradable and/or toxic organic pollutant (Almomani et al, 2010, Al

Momani and Jarrah, 2009). Using solar irradiation as a source of photon can reduce the cost of AOPs and make them a suitable treatment options.

The present work aims to investigate the suitability of using solar induced AOTs for removal of ECs from water and wastewater. The study will be carried out in a compound parabolic collector (CPC) prototype pilot plant solar reactor. The CPC technology allows all the received solar radiation to be utilized for oxidation. Thus, oxidation performance increase. The outcome of this project will allow to develop a new technology for the treatment of water and wastewater containing ECs by use of solar energy which is abundant in Qatar and the Arab gulf region.

Materials and Methods

Experimental Set Up

Experiments set up used to carry out pharmaceutical solar oxidation is shown in (Fig. 1). The solar pilot plant is equipped with (1) a wastewater containing pharmaceutical. The tank was immersed in water bath to control the working temperature, (2) compound parabolic collector (CPC) solar reactor . the CPC allows for almost all the solar irradiation received at the reactor aperture area (not only direct, but also diffuse) to be collected being available for oxidation. The solar consists of six quartz tubes with a length of 50 cm, an inner diameter of 3 cm and outer diameter of 3.1 cm. This pilot plant has the capacity to treat up 5 L of wastewater, the solution was circulated through the reactor with a flow rate of 20 L. h^{-1} and the sun's movement was tracked manually.

Analytical methods

The pharmaceutical concentrations were measured with a Hewlett-Packard 1050 series HPLC equipped with an Ultra Aqueous C_{18} column from Restek (3.2100 mm, 5 mm) and a variable wavelength detector. The eluents consisted of 10 mM phosphoric acid (A) and pure acetonitrile (B). The eluent flow rate was 0.5mL.min^{-1} and the detection limit was 10 μg.L^{-1}. The 95% confidence intervals for a single measurement were typically 3–5%.

The dissolved organic carbon (DOC) was measured using a Shimadzu TOC analyzer. The analyzer was calibrated using potassium phthalate standard solution and following the instrument standard method. Each sample was analyzed three times and measurements were averaged. Measurements reproducibility showed a precision in the range of ±0.1 mg.C.L^{-1}.

Water samples

Wastewater (WW) was collected from the effluent of secondary treatment train from a Doha north WWTP. Water samples were filtered through a 0.45 μm Millipore filter, and stored at 4 °C prior to use. Table 1 shows the characteristics of Wastewater treatment plant effluent.

Table 2: Water quality parameters for surface and wastewaters.

Parameter	Wastewater treatment plant effluent Average, (std. dev.)[1],
pH	7.9 (0.4)
DOC (mg.L^{-1})	13(1.2)
Alkalinity (mg$CaCO_3$.L^{-1})	145 (6.0)
Nitrate (mg-N.L^{-1})	31 (1.4)

Nitrite (mg-N.L^{-1})	15 (0.4)
Sulfate (mg.L^{-1})	50 (2.0)
Chloride (mg.L^{-1})	15 (1.2)

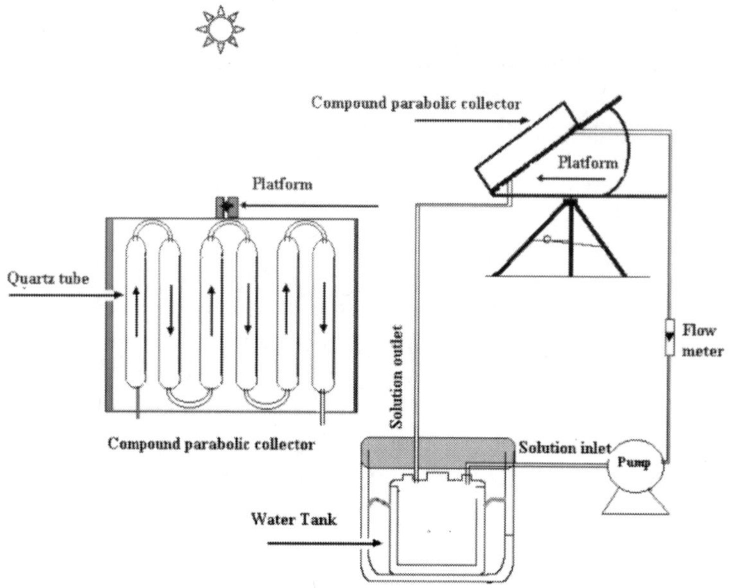

Figure 1: Solar Experimental Setup

Experimental Procedure

Experiments were conducted Doha, Qatar (25.2854° N, 51.5310° E) in a semi-batch mode, under sunlight and at ambient temperature. At the beginning of the reactor was loaded with 5 L of the aqueous solution of pharmaceuticals and the required amount of catalyst was added afterwards. Then, the solution was recirculated through the CPC by a centrifugal pump at a flow rate of 20 L.hr^{-1} so that the flux pattern inside the tubes was laminar (the Reynolds number was 2500). For the first 10 min, the CPC tubes were kept covered to avoid sunlight illumination. This time was enough to reach the adsorption equilibria of pharmaceuticals onto TiO_2 (if used). After that, the CPC was uncovered and the reaction was was started. Also, if needed, hydrogen peroxide was added at this point. Samples were withdrawn at intervals through a sampling port to determine the concentration of pharmaceuticals as function of reaction time, the remaining total organic carbon (TOC), and the concentrations of hydrogen peroxide, ozone and Fe(II). The ozone concentration in the outlet gas stream was also continuously monitored with in-line ozone analyzers.

The initial concentration of pharmaceuticals used in experiment are presented in table 1. The initial concentrations were selected to improve the chemical analysis accuracy and facilitate mineralization kinetic study. The initial Fe(III) and TiO_2 concentrations were 3 and 100 mg L^{-1}, respectively.

Toxicity analysis

Toxicity bioassays were performed to evaluate the toxicity of water containing pharmaceuticals before and after ozonation. Tests were carried out using Microtox® bioassays, which involved the use of the marine bacterium *Photobacterium phosphoreum* as an indication of sample toxicity. The standard Microtox protocol was followed, and data analysis was achieved using the Microtox™ Omni Software (SDI, 2002). Sample toxicity was evaluated at 5 and 15 min from the time of mixing of the sample with marine bacterium at various dilutions. The toxicity was determined by measuring the EC_{50} of the sample. The EC_{50} is the concentration of the test sample at which there is 50% reduction in bacterial light production. This value was related to sample toxicity using the following equation:

$$TU = \frac{100}{EC_{50}} \qquad [1]$$

where *TU* is the toxicity unit and EC_{50} is the concentration of the test sample at 50% reduction in bacterial light production.

Results and Discussion

***Solar -Photolysis of* pharmaceuticals**

Figure 2 A and B show the evaluation of pharmaceutical as a function of solar radiation time during solar photolysis process. Direct solar photolysis was not effective in removing or degradation of pharmaceuticals from wastewater. This is an expected fact as the solar radiation with wavelength < 290 nm does not reach Earth's surface as it is mostly retained by atmosphere [32]. In addition, all the target pharmaceuticals do not absorb radiation above 300 nm significantly. The removal efficiencies ranged from 9.1- 12.1%, 13-15.6%, 13.9 -16 and 5.4 and 8.2 for antibiotics, estrogen, acidic and neutral pharmaceutical, respectively.

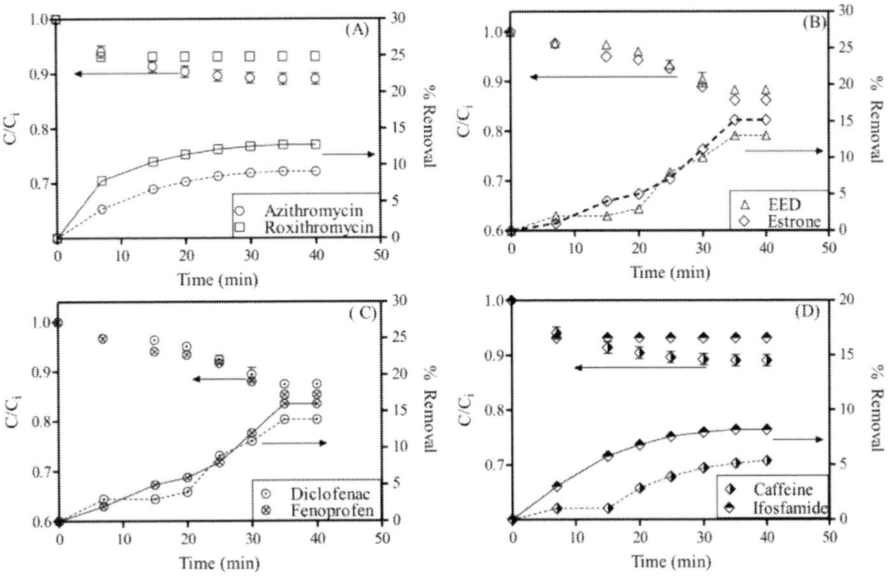

Figure 2: Evolution of the normalized concentration $[C]/[C]_i$ and % removal of antibiotics, Estrogens, Acidic and natural pharmaceuticals during solar photolysis

Solar-photocatalytic oxidation of pharmaceuticals

Iron (Fe(III) and titanium dioxide were used as catalyst for this part of experiments. Figure 2 shows the results for experiments carried out with iron as catalyst in the AOPs including Fe(III)/solar-hv and Fe(III)/solar-hv/H_2O_2. The reactions involved in these oxidation processes are summarized in equations (1 and 2).

$$Fe^{3+} + H_2O \rightleftarrows Fe(OH)^{2+} + H^+ \qquad \text{(Reaction 1)}$$

$$Fe(OH)^{2+} \xrightarrow{h\nu} Fe^{2+} + {\cdot}OH$$ (Reaction 1)

During the photocatalytic oxidation with Fe(III)/hv Pharmaceuticals oxidation occurs by the formation of aqua complexes, mainly $Fe(OH)_2^+$ (see reaction (1)), which can undergoes photolysis under UV radiation ($\lambda > 300$ nm) to generate hydroxyl radicals as shown reaction 2. As it can be seen in Figure 3, the combination Fe(III)/hv led to reasonable degradation of the pharmaceuticals. The removal efficiencies for antibiotics, estrogen, acidic and neutral pharmaceutical ranged from 74 – 77.8%, 69-71.9%, 70.8-80.9% and 57.6-66.8, % respectively. The results suggest that solar-photocatalytic oxidation can be used to oxidize these pharmaceuticals. However, the reaction time should be increase of there is a need to add reagent to assist in oxidation.

Adding H_2O_2 to Fe(III)/hv change the system a well-known oxidation system called photo-Fenton. In this case, H_2O_2 accelerates the production of hydroxyl radicals as in reactions (3) and (4) and change Fe (III) to Fe(II) (reaction 5) which, in turn, can reduce Fe(III) to Fe(II) as in reaction (6). The net results is more hydroxyl radicals and more pharmaceuticals oxidation.

$$Fe^{2+} + H_2O_2 \rightarrow Fe^{3+} + OH^- + {\cdot}OH$$ (Reaction 3)

$$H_2O_2 \xrightarrow{h\nu} 2{\cdot}OH$$ (Reaction 4)

$$H_2O_2 + {\cdot}OH \rightarrow HO_2^{\cdot} + H_2O$$ (Reaction 5)

$$Fe^{3+} + HO_2^{\cdot} \rightarrow O_2 + Fe^{2+} + H^+$$ (Reaction 6)

Solar photo catalytic oxidation by TiO_2 is well-known and it has been widely used to oxidize non-biodegradable organic matter in water systems.

The change in solution toxicity after different oxidation process were evaluated. For experiments carried out with photolysis, it was observed that oxidized water samples toxicity does not change a suggesting that the low oxidation reaction did not change water toxicity. For water samples with higher oxidation such as photo-Fenton and TiO_2/hv system lower acute toxicity than the initial wastewater was observed. As an example, water containing 3.75 mg/L of Roxithromycin showed a TU value of 17, oxidation photo-Fenton and TiO_2/hv system decreased the TU to 1 and 1.2, Respectively. Similar results were observed for other pharmaceutical. The reduce in toxicity is related to the reaction between radicals and the studied pharmaceuticals at tertiary amines bonds leading to the cleavage of the N–C bonds. This reaction can lead to either the destruction of the chemical structure of pharmaceuticals or the loss of the methyl group from the amino group. Li et al., 2001 have confirmed that the loss of the methyl group of roxithromycin eliminates its apparent activity. Besides, the oxidation reaction can cause cleavage of the phenolic structure in pharmaceuticals and reduce its estrogenicity significantly. On the basis of these considerations, solar photocatalytic oxidation of pharmaceuticals results in significant reduction in pharmacological effects of these pharmaceuticals

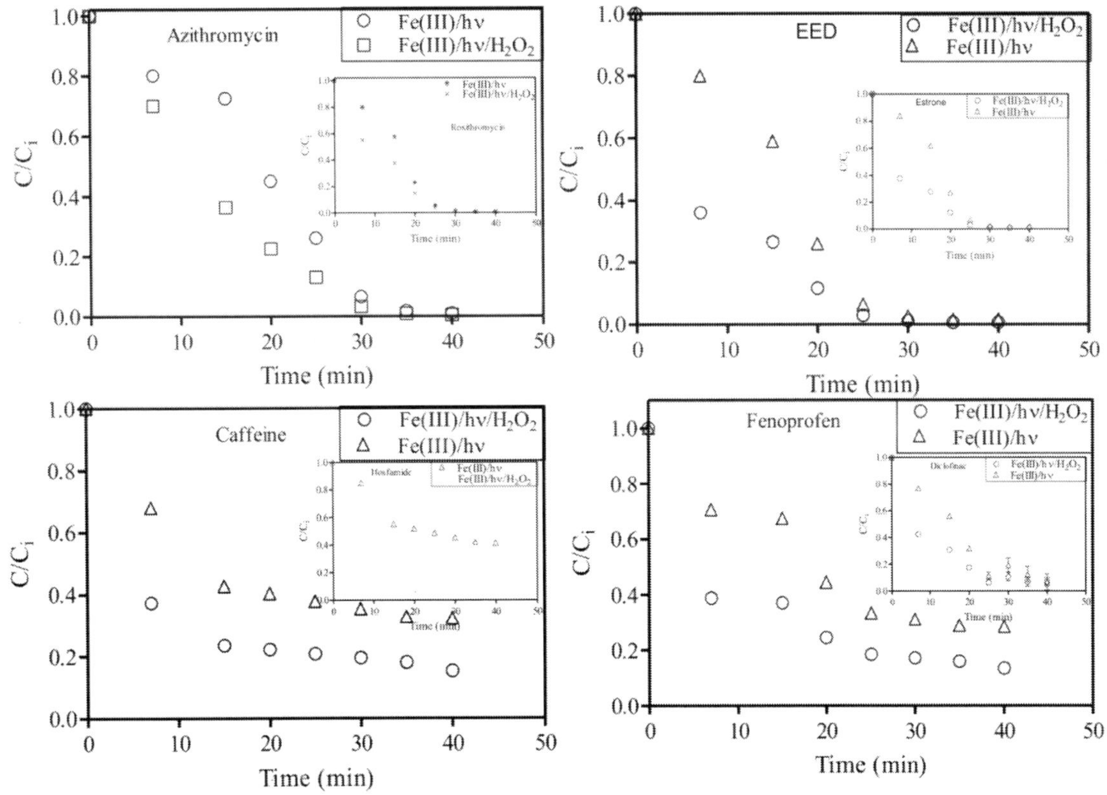

Figures 3 : Evolution of the normalized concentration $[C]/[C]_i$ of antibiotics, Estrogens, Acidic and natural pharmaceuticals during solar-photocatalytic oxidation.

Conclusion:

Oxidation of four groups of pharmaceuticals (antibiotics, estrogens, acidic and neutral) with solar advanced oxiadation was investigated in effluent of WWTP. Solar photocatalytic reaction can be used successfully to eliminates nob-biodegradablke pharamacuticals from WWTP effluets. The degradation rate of pharmaceuticals was strongly accelerated by adding catalys or hydrogen peroxide to the system. 98 % of pharmaceutical degradation was achieved by solar photo-Fenton or photo-catakytic oxidation with TiO_2in 40 min reaction time.
Toxicity tests using Microtox™ bioassay were completed to evaluate the effect of the oxidation reactor on the toxicity of water solution before and after the treatment. Solar oxidation reduced the a cute toxicity of all the pharmaceuticals in the wastewater suggesting that this technology can be coupled with wastewater treanment process to complete the removal of the toxic and non-biodegradable materials

Acknowledgements
This publication was made possible by the NPRP grant (NPRP No.: 6 - 1436 - 2 - 581) from the Qatar National Research Fund (a member of Qatar Foundation). The statements made herein are solely the

responsibility of authors. The authors gratefully acknowledge the support provided by the Qatar University to do this work..

REFERENCES

DODD, M. C. & HUANG, C.-H. (2004) Transformation of the Antibacterial Agent Sulfamethoxazole in Reactions with Chlorine: Kinetics, Mechanisms, and Pathways. *Environmental Science & Technology,* 38, 5607-5615.

GAMAL EL-DIN, M., SMITH, D. W., AL MOMANI, F. & WANG, W. (2006) Oxidation of resin and fatty acids by ozone: Kinetics and toxicity study. *Water Research,* 40, 392-400.

HUBER, M. M., KORHONEN, S., TERNES, T. A. & VON GUNTEN, U. (2005) Oxidation of pharmaceuticals during water treatment with chlorine dioxide. *Water Research,* 39, 3607-3617.

HUBER, M. M., TERNES, T. A. & VON GUNTEN, U. (2004) Removal of Estrogenic Activity and Formation of Oxidation Products during Ozonation of 17alpha -Ethinylestradiol. *Environmental Science & Technology,* 38, 5177-5186.

KOLPIN, D. W., FURLONG, E. T., MEYER, M. T., THURMAN, E. M., ZAUGG, S. D., BARBER, L. B. & BUXTON, H. T. (2002) Pharmaceuticals, Hormones, and Other Organic Wastewater Contaminants in U.S. Streams, 1999-2000: A National Reconnaissance. *Environmental Science & Technology,* 36, 1202-1211.

LOVATO, M. A. E., MARTÃ-N, C. A. & CASSANO, A. E. (2009) A reaction kinetic model for ozone decomposition in aqueous media valid for neutral and acidic pH. *Chemical Engineering Journal,* 146, 486-497.

PINKSTON, K. E. & SEDLAK, D. L. (2004) Transformation of Aromatic Ether- and Amine-Containing Pharmaceuticals during Chlorine Disinfection. *Environmental Science & Technology,* 38, 4019-4025.

REDDERSEN, K., HEBERER, T. & DANNBIER, U. (2002) Identification and significance of phenazone drugs and their metabolites in ground- and drinking water. *Chemosphere,* 49, 539-544.

ROSENBLATT, D. H., HULL, L. A., DE LUCA, D. C., DAVIS, G. T., WEGLEIN, R. C. & WILLIAMS, H. K. R. (1967) Oxidations of Amines. II. Substituent Effects in Chlorine Dioxide Oxidations. *Journal of the American Chemical Society,* 89, 1158-1163.

TERNES, T. (2001) Pharmaceuticals and Metabolites as Contaminants of the Aquatic Environment. *Pharmaceuticals and Care Products in the Environment.* American Chemical Society.

TERNES, T. A., MEISENHEIMER, M., MCDOWELL, D., SACHER, F., BRAUCH, H.-J. R., HAIST-GULDE, B., PREUSS, G., WILME, U. & ZULEI-SEIBERT, N. (2002) Removal of Pharmaceuticals during Drinking Water Treatment. *Environmental Science & Technology,* 36, 3855-3863.

WEBB, S., TERNES, T., GIBERT, M. & OLEJNICZAK, K. (2003) Indirect human exposure to pharmaceuticals via drinking water. *Toxicology Letters,* 142, 157-167.

ZHANG, P. & LIU, J. (2004) Photocatalytic degradation of trace hexane in the gas phase with and without ozone addition: kinetic study. *Journal of Photochemistry and Photobiology A: Chemistry,* 167, 87-94.

481851 Oil Pollution, Environmental Degradation and Restoration of Nigeria's Ogoniland

Abstract:

Human activities especially those of oil exploration and exploitation raise a number of issues such as depletion of biodiversity, coastal and riverbank erosion, flooding, oil spillage, gas flaring, noise pollution, sewage and wastewater pollution, land degradation, soil fertility loss and deforestation, which are all major environmental issues.

Ogoniland is characterized by typically deltaic features; uneven terrain, numerous creeks, shallow brackish water bodies and a variety of vegetation types including swamp forest. The land has a groundwater either as localized aquifers or in hydraulically interconnected aquifers. Ogoniland is not an Island. This has two implications. The first is that pollution from Ogoniland has the potential to reach and cross its boundaries as well as entering Ogoniland from extreme sources.

Petroleum hydrocarbons are naturally occurring hydrocarbons substances and depending on the length of the carbon, can occur in gas, liquid and solid form. However, hydrocarbons refer to chemical substances formed exclusively from carbon and hydrogen by the decay of organic substances trapped within sedimentary rocks.

Hydrocarbons substantially alter the physical and chemical properties of the soil once it comes in contact with it. The contact could be from the natural seepage of hydrocarbons to accidental seepage of crude oil on the ground. In the case of water, hydrocarbons can cause both physical and chemical effects by preventing oxygen transfer in the water column, thus affecting aquatic life-support systems. As regards vegetation, where spillage is not immediately attended to, oil spills often lead to fires, causing total or partial destruction of vegetation. Dissolved or emulsified oil in the water column can contaminate the plankton, algae, fish, eggs and the vertebrae larvae. Also, physical contacts with oil destroys the insulation of fur and feathers, and more often, birds ingest oil which have lethal or sub-lethal impacts through, for example, liver and kidney damage.

Residents in these areas, face the risk of dangerous exposure to hydrocarbons when they breathe, bathe, eat fish, drink water or accidentally eat or touch soil or sediment that is contaminated with oil. In addition to the chemical pollution by hydrocarbons, there are other environmental concerns linked with oil industry operations. These range from clearance of land for oilfield facilities, hydrological changes due to construction of roads and pipelines and contamination from chemicals other than hydrocarbons.

There must be maintenance of oilfield facilities, decommissioning of oilfield facilities and prevention of illegal activities, oil spill response and remediation of contaminated sites. Environmental restoration includes the establishment of an Integrated Contaminated Soil Management Centre (ICSMCS), setting up of mini-treatment centers and rehabilitation of mangroves.

Part of the actions is the amendment of the official gazette establishing the Hydrocarbon Pollution Restoration Project (HYPREP), to reflect a new governance framework comprising a Governing Council, a Board of Trustees and Project Management.

There must be an end to the widespread pipeline sabotage, crude oil theft and illegal refining that are the main cause of environmental damage in Ogoniland and the wider Niger Delta.

There should be transfer of oversight of the Environmental Guidelines and Standards for Petroleum Industry in Nigeria (EGSPIN), to ensure constant surveillance. Also, there should be the provision for social and health impact assessment as an integral part of the overall environmental impact assessment (EIA) process for all new oil and gas facilities and upgrades to existing facilities in line with international best practice. This means that the Environmental Impact Assessment (EIA) Act should be amended to accommodate these changes. It also proposes that there should be a review of the provisions of the National Oil Spill Detection and Response Agency NOSDRA (Establishment) Act, 2006 against NOSDRA's current operational responsibilities. Oversight of clean-up should be given to a separate governmental department.

POTENTIAL USE OF MIXED INDIGENOUS MICROALGAE FOR CARBON DIOXIDE BIO-FIXATION AND ADVANCED WASTEWATER TREATMENT

Fares Almomani[1], Simon Judd[2], Mohamad Shurai[1], Rahul Bhosale[1], Anand Kumar[1], Majeda Khreisheh[1]

1) College of Engineering, Department of Chemical Engineering, Qatar University, P. O. Box 2713, Doha, Qatar

2) Gas Processing Center, Qatar University; 2Department of Chemical Engineering, Qatar University

ABSTRACT

The potential use of mixed indigenous microalgae (MIMA) as a CO2 capturing technology and treatment process for wastewaters was evaluated at two temperatures (25 and 30 °C) and different CO2 dosage (0%, 5% and 15%). During the study the growth rate of MIMA, CO_2 utilization capacity, removals of organic matter and nutrients from synthetic wastewater were followed. A noticeable difference between the growth patterns of MIMA was observed at different CO_2 and different operational temperatures. MIMA showed the highest growth grate when injected with CO_2 dosage of 5% compared to blank solution 0% CO_2 injection. Ammonia removals for MIMA were 94%, 93%, and 84% for the media injected with 0, 5 and 10% CO_2 at 25°C. The results of this study show that simple and cost-effective microalgae-based wastewater treatment systems can be successfully employed at different temperatures as a successful CO_2 capturing technology.

INTRODUCTION

In the near future, improving the overall efficiency of the Carbon Capture and Storage (CCS) chain will become increasingly important, as the implementation of the various CO_2 existing capture methods from flue gas are not currently economically feasible. On the other hand, it is to be expected that CO_2 will become available from a multitude of different sources. For the GCC (Gulf Cooperation Council) this will include power plants, refineries, various petrochemical processes (methanol, ammonia), and will encompass gas separation at processing plants and at LNG production facilities. CO_2 has to be removed from flue gas (CO_2-N_2) and from natural gas (CO_2-CH_4) mixtures.

The current average costs of CO_2 capture and storage, based on chemical absorption, are estimated to be at least 55 US$ per ton of CO_2 taking a coal-fired power plant as a reference. In the absorption process, the solvent regeneration step accounts for approximately 75% of the energy consumption (and thus costs). The high energy penalty, which for power plants can be as high as 10% (based on the overall power plant efficiency of 35-45%), is one of the main drivers for research into more economically viable, and ultimately sustainable, approaches to be included in the full CCS chain.

One of the emerging approaches is based on so-called CO_2 utilization, in which CO_2 is considered as a feedstock for, for example, biological systems. Of such approaches the one which has garnered the most interest is the use of micro-algae for the biofixation of CO_2, in combination with the production of a range of bioproducts. In a number of papers an overview is given of the amount of CO_2 that can converted by different types of algae. The main way to identify suitable algae is based on the so-called CO_2 fixation rate, the amount of CO_2 converted in mg (milligram) of CO_2 per day per unit of volume (in liters).

At the same time, with the increase on world population the high volumes of wastewater generated by human was always a permanent problem that requires appropriate treatment. While domestic wastewater treatment plants (WWTPs) is considered an effective treatment for removal of biochemical oxygen demand (BOD), suspended solids and bacteria, the removal of inorganic nutrients, still limited and challenging in these processes. With the implementation of the new wastewater discharge regulation, the removal of nutrients, mainly dissolved nitrogen and phosphorus, is becoming one of the essential requirement for the approval of the WWTP performance. The new regulations were based on the fact that the presence of these nutrients in wastewater will leads to eutrophication by stimulating the growth of unwanted plants such as algae and aquatic macrophytes, and increase the effluents toxicity to fishes and aquatic organisms. Moreover, untreated nutrients in wastewater effluent reduce the performance of disinfection step and increase the chlorine demand. As a result, there is a big need to find a treatment process that can remove these nutrients before the effluents are discharged.

So far, limited attention has been paid to the dual role of microalgae as advanced technology for treatment of different kinds of wastewater and as CO_2 bio-fixation technology. Moreover, there are knowledge gaps in understanding the performance of different microalgae strains in the treatment of wastewater and their capacity for CO_2 capturing. In particular, according to our literature review, no studies were carried out using mixed indigenous microalgae cultures as opposed to lab-grown pure culture strains. This knowledge is needed to select and match the right type of microalgae with the right type of wastewater, which is crucial in improving the operation and performance of microalgae-CO_2-biofixiation- wastewater treatment systems

This study evaluated the use of mixed indigenous microalgae (MIMA) as carbon dioxide bio-fixation technology and advanced wastewater treatment. The study follows the growth rate of MIMA, removals of organic matter, removal of nutrients from wastewater and its effectiveness as CO_2 capturing technology from flue gas.

CULTURE SYSTEM

Algae growth experiments were carried out in 1 L- sterilized Erlenmeyer flasks at 36°C. The growth experiments were performed using an incubator equipped with thermostat and two cool white lamps of 1500 lux each. Temperature control was maintained at 8, 20 and 36 °C by a fan connected to thermostatic control. Two fluorescent —daylight lamps were used as light sources. The photon flux intensity inside the incubator was measured by NIST Traccable Radiometer (International light. Model IL 400A) and found to be 1.5 μW/cm2.

A volume of 850 mL of wastewater was mixed with algal strain and incubated at constant temperature and continuous lighting for a period of 28 days.

DETERMINATION OF ALGAL GROWTH

Samples were taken from the cultivation media every day for measurement of optical density at 690 nm (OD_{690}) using a spectrophotometer (VARIAN 100 Bio UV-visible spectrophotometer) as the algal density indicator. The growth rate (GR, per day) was calculated as follows:

$$Growth\ rate\ (GR) = \frac{\ln(OD_{690nm,t}) - \ln(OD_{690nm,0})}{\Delta t} \qquad (1)$$

where $OD_{690nm,0}$ is the optical density on the initial day, $OD_{690nm,\ t}$ is the optical density on a selected day. to eliminate the interference of the background wastewater, each recorded OD was corrected by taking away that of the OD corresponding blank sample (original wastewater).

RESULTS AND DISCUSSION

Algae Growth at different temperatures

Algal growths of mixed indigenous microalgae (MIMA) culture in terms of optical density (OD) at 690 nm 25 ◦C) and different CO_2 dosage were plotted as function of time in Figure 1. The growth curve of cultivated algae follows a typical growth curves consisting of lag phase, exponential growth phase and stationary phase. It can be seen in the figure 1 that the lag time for MIMA range from 1 to 5 days, and the best growth trend was obtained for gas stream containing 5 % CO2. The measured OD was converted into dry cell weight using pre-determined conversion factors. The final biomass concentrations for experiments carried out with MIMA after 24 days of cultivation were found to be 1.31, 1.71 and 1.27 g/L for system operated with 0, 5 and 10 % CO_2, respectively.

The average specific GR and biomass productivity of MIMA at different CO2 injection were determined. The average specific growth rates in the first 3 days of growth phase (days 3, 4 and 5) were found to be 0.22, 0.20, and 0.40 day^{-1} for for system operated with 0, 5 and 10 % CO_2, respectively.

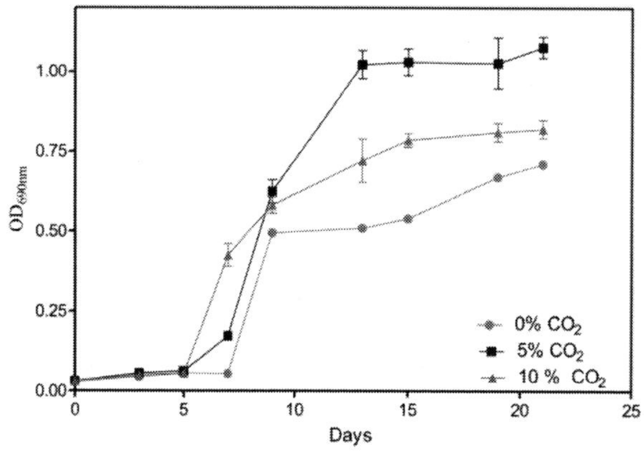

Figure 1: The change in optical density of MIMA in as a function of CO_2 dosage and cultivation time

Organic matter removals

Figure 2 shows the removal of COD as function of cultivation time for MIMA in synthetic WW and at different CO2 dosage. COD showed a rapid decrease in the first few weeks which then stabilized and remained same until the end of the cultivation time. COD removal efficiency also varied among different wastewaters.. Rapid decrease in COD was achieved during the first 7 days of cultivation for all wastewater samples. Significant COD removals ls were achieved with the mixed indigenous culture injected by 5% CO_2.

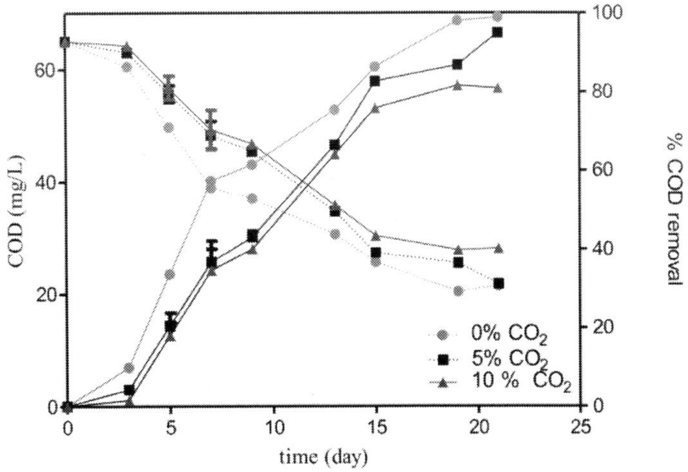

Figure 2: COD removal as a function of CO_2 dosage and cultivation time.

Removal of ammonia

MIMA showed very good efficiencies in removing ammonia (NH_4–N) from wastewater (Figure 3). The removal efficiencies were 94%%, 93%% and 84% for system operated with 0, 5 and 10 % CO_2, respectively.

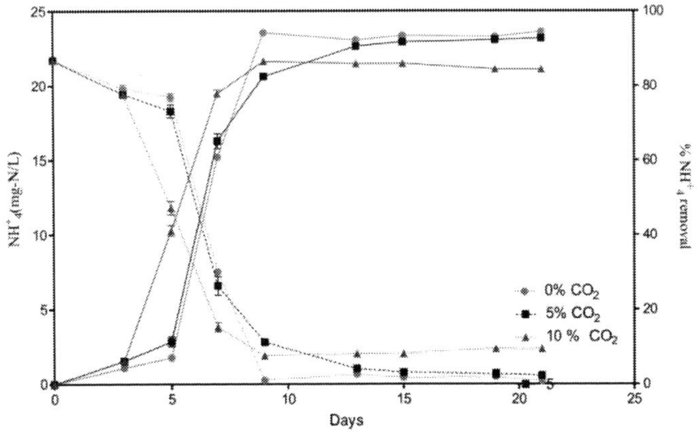

Figure 3: NH_4^+ removal by MIMA as a function of CO_2 dosage and cultivation time.

Removal of phosphorus

Figure 4 shows the evaluation and percent removal of total dissolved phosphorus for system operated with 0, 5 and 10 % CO_2 with mixed microalgae culture. The general trend indicated that the highest phosphorus removal was achieved with system injected with 0% CO_2. Experiments carried out with Mixed indigenous microalgae showed higher phosphorus removal efficiencies of 95%, 93% and 84% were achieved for the solutions injected with 0, 5 and 10 % CO_2, respectively.

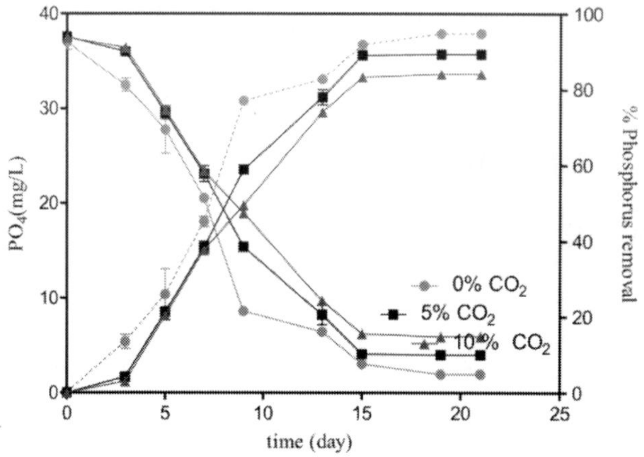

Conclusions

The results of this study showed that *MIMA is* effective in capturing CO_2 and removing ammonia and phosphorus from wastewater at 25 and 30°C. This simple and cost-effective microalgae-based wastewater treatment systems can be successfully employed as a simple and cheap CO2 capturing technology in combination with wastewater treatment. *MIMA* showed a maximum growth for media injected with 5% of CO_2 at 25 and 30◦C. More than 90% of ammonia and 90 % of Phosphorous was removed by *MIMA* in the temperature range of 20-30°C. Growth rates of microalgae were different not only between the two temperatures but also for the same microalgae injected with different CO_2, which indicates that the operational condition play an important role in determining the potential of algae as CO_2-capturing technology and this knowledge can be used to optimize the microalgae based treatment systems in hot regions.

ACKNOWLEDGMENT

This publication was made possible by the NPRP grant (NPRP No.: 6 - 1436 - 2 - 581) from the Qatar National Research Fund (a member of Qatar Foundation). The statements made herein are solely the responsibility of authors. The authors gratefully acknowledge the support provided by the Qatar University to do this work.

REFERENCES

Abdel-Rauf, N., Al-Homaiden, A. A., and Ibraheem, I. B. M. 2012. Microalgae and wastewater treatment. Saudi Journal of Biological Sciences, **19:** 257-275.

APHA, A., WEF, editor. 1995. Standard Methods for the Examination of Water and Wastewater. 19 edition.
Hamilton, S. K., and J. W.M. Lewis. 1990. Basin morphology in relation to chemical and ecological characteristics of lakes on the Orinoco River floodplain, Venezuela. Archiv für Hydrobiologie **119**:393-425.
Hamilton, S. K., S. J. Sippel, D. F. Calheiros, and J. M. Melack. 1997. An anoxic event and other biogeochemical effects of the Pantanal wetland on the Paraguay River. Limnology and Oceanography **42**:257-272.
Li, W. K. W. 1980. Temperature adaptation in phytoplankton: Cellular and photosynthtic characteristics. In: Falkowski, P. G. (ed.) Primary productivity in the sea. Plenum Press.
Li, Y., M. Horsman, B. Wang, N. Wu, and C. Q. Lan. 2008. Effects of nitrogen sources on cell growth and lipid accumulation of green alga Neochloris oleoabundans. Applied Microbiology and Biotechnology **81**:629 – 636.
Mahapatra, D. M., Chanakya, H. N., Ramachandra, T. V., 2013. Treatment efficacy of algae-based sewage treatment plants. Environmental Monitoring and Assessment, **185,**7145-7164.
Martin, C., De la Noae, J., Picard, G., 1985. Intensive cultivation of freshwater microalgae on aerated pig manure. Biomass, **7,** 245-259.
Melo, S. d., and V. L. M. Huszar. 2000. Phytoplankton in an Amazonian flood-plain lake (Lago Batata, Brasil): Diel variation and species strategies. Journal of Plankton Research **22:**63-76.
Moheimani, N. R. and Borowitzka, M. A. 2006. Limits to productivity of the algae *Pleurochrysis carterae* (haptophyta) grown in outdoor raceway ponds. Biotechnology and Bioengineering **96 (1)**: 27-36.
Mulbry, W., Kondrad, S., Pizarro, C., Kebede-Westhead, E., 2008. Treatment of dairy manure effluent using freshwater algae: Algal productivity and recovery of manure nutrients using pilot-scale algal turf scrubbers. Bioresource Technology, **99,** 8137-8142.
Park, J. B. K., Craggs, R. J., and Shilton, A. N. 2011. Wastewater treatment high rate algal ponds for biofuel production. Bioresource Technology, **102:**35-42.
Pittman, J. K., Dean, A. P., and Osundeko, O. 2011. The potential of sustainable algal biofuel production using wastewater resources. Bioresource Technology, **102:** 17-25.
Pittman, J. K., Dean, A. P., Osundeko, O., 2011. The potential of sustainable algal biofuel production using wastewater resource. Bioresource Techn, **102,** 17-25.
Pouliot, Y., Buelna, G., Racine, C., De la Noüe, J., 1989. Culture of cyanobacteria for tertiary wastewater treatment and biomass production. Biological Wastes, **29,**81-91.
Ratkowsky, D. A., Lowry, R. K., McMeekin, T. A., Stokes, A. N. and Chandler, R. E. (1983) Model for bacterial culture growth rate throughout the entire biokinetic temperature range. J. Bacteriology **154:** 1222-1226.
Salvucci, M. E., Crafts-Brandner, S. J. 2004. Inhibition of photosynthesis by heat stress: the activation state of Rubisco as a limiting factor in photosynthesis. Physiology Plant **120(2)**: 179-186.
Sutherland, D. L., Montemezzani, V., Howard-Williams, C., Turnbull, M. H., Broady, P. A., Craggs, R. J., 2015. Modifying the high rate algal pond light environment and its effects on light absorption and photosynthesis. Water Research, **70,** 86-96.
Tarlan, E., Dilek, F. B., Yetis, U., 2002. Effectiveness of algae in the treatment of a wood-based pulp and paper industry wastewater. Bioresource Technology, **84,** 1-5.
Van Den Hende, S., Beelen, V., Bore, G. I., Boon, N., Vervaeren, H., 2014. Up-scaling aquaculture wastewater treatment by microalgal bacterial flocs: From lab reactors to an outdoor raceway pond. Bioresource Technology, **159,** 342-354.
Wang, B. and C. Q. Lan. 2011. Biomass production and nitrogen and phosphorus removal by the green alga Neochloris oleoabundans in simulated wastewater and secondary municipal wastewater effluent. Bioresource Technology **102**:5639-5644.

Wang, L., M. Min, Y. Li, P. Chen, Y. Chen, Y. Liu, Y. Wang, and R. Ruan. 2009. Cultivation of Green Algae Chlorella sp. in Different Wastewaters from Municipal Wastewater Treatment Plant. Applied Biochemistry and Biotechnology **162**:1174-1186.

Wang, L., Min, M., Li, Y., Chen, P., Chen, Y., Liu, Y., Wang, Y., Ruan, R., 2009. Cultivation of Green Algae Chlorella sp. in Different Wastewaters from Municipal Wastewater Treatment Plant. Applied Biochemistry and Biotechnology, **162,** 1174-1186.

Woertz, I. C., Fulton, L., Lundquist, T. J., 2009. Nutrient Removal & Greenhouse Gas Abatement with CO2 Supplemented Algal High Rate Ponds.*in* W. E. Federation, editor. WEFTEC annual conference, Orlando, Florida.

Zamalloa, C., Boon, N., Verstraete, W., 2013. Decentralized two-stage sewage treatment by chemicalâ€"biological flocculation combined with microalgae biofilm for nutrient immobilization in a roof installed parallel plate reactor. Bioresource Technology, **130,** 152-160.

Zamalloa, C., Vulsteke, E., Albrecht, J., Verstraete, W., 2011. The techno-economic potential of renewable energy through the anaerobic digestion of microalgae. Bioresource Technology, **102,** 1149-1158.

479981 Modeling of Simultaneous Absorption of CO_2 and NH_3 in a Type I Turbulent Contact Absorber

Abstract:

A turbulent contact absorber (TCA), also named turbulent bed contactor or mobile bed contactor, is a countercurrent gas–liquid–solid fluidized bed in which liquid flows down through a turbulent bed of inert and spherical particles supported by continuous gas phase. TCA with low pressure drop and high contact efficiency has the advantages of enhanced heat and mass transfer rate, non-choking and self-cleaning owing to the tumbling motion of the packing. This makes the TCA more suitable over packed or tray columns for many physical, chemical and biochemical processes especially for CO_2 capture form coal-fired flue gas. However, the difference of the countercurrent gas–liquid contact structure operating at different flow regimes of a TCA significantly affects the interfacial area and mass transfer coefficients calculation. An accurate prediction of the mass transfer coefficients is very important for the design and optimization of TCA.

- This article presents the modeling and experimental investigation of the simultaneous absorption of CO_2 and NH_3 into water in a TCA.
- A theoretical model is developed to describe the gas-liquid countercurrent flow characteristic and interfacial area for the prediction of the mass transfer coefficients K_Ga in a type I TCA with fully fluidization operation.
- The dispersed liquid flows in the TCA are assumed to be comprised of film flows adhere to the packing hollow balls and droplets flows in the void of the fluidization bed.The overall gas-side mass transfer area consists of the mass transfer coefficient of the packing and mass transfer coefficient of the droplets.
- The thickness of the film to determine the liquid holdup adhere to the packing balls is modeled as applying the shear stress boundary condition at the liquid-gas interface and the non-slip boundary condition at the sphere surface.
- The saunter mean diameter of droplets is obtained as based on the balance relationship between total surface free energy of liquid droplets and turbulent kinetic energy of gas flow based on the locally isotropic turbulence of Kolmogorov.
- The concentration profile equations and mass transfer coefficients of CO_2 absorption in the droplet and in the film of the TCA are analyzed respectivelly.
- Model reliability is verified by comparing the model results with the experimental investigation of simultaneous absorption of NH_3 and CO_2 into water in a fully fluidized TCA under the superficial liquid velocity of 0.001~0.005m/s and superficial gas velocity of 1.24~2.07m/s with different packing diameters of 0.01~0.025m and static packing heights of 0.1~0.25m. The theoretical overall volumetric mass transfer coefficients are calculated to be $5\times10^{-5}\sim8\times10^{-5}mols^{-1}m^{-3}Pa^{-1}$ including the gas-liquid mass transfer across the droplets and liquid film adhere to the packing balls.The model prediction compared reasonably to the experimental results with the maximum deviation of approximately 27%.

480976 CO_2 Conversion to Methanol through Bi-Reforming

Abstract:

The technology options available to convert CO_2 into higher value fuels and chemicals are not commercially well established. Developing commercially viable technologies to convert CO_2 into high value products will improve the project economics of renewable and non-renewable carbon-based energy production. Such technologies can potentially create additional revenue streams with significant reduction in the net CO_2 emissions. A well-known pathway for CO_2 utilization is reforming of CH_4 in presence of steam and CO_2 (bi-reforming of CH_4) for syngas ($CO+H_2$) production. The syngas can then be used as a raw material for synthesis of fuels and chemicals, including methanol. The specific reactant stoichiometric ratio of CH_4/steam/CO_2 of 3/2/1 creates the syngas ratio of 2 which is critical for methanol synthesis. In this context, an Aspen plus process simulation model for bi-reforming reaction system coupled with the methanol synthesis reactor system is developed. The model considers a range of operating conditions and is used to optimize the integrated process. Under optimum operating conditions, the overall process thermal efficiency is 65%. The results from process simulation are used to conduct a Life Cycle Analysis (LCA) of the proposed pathway. Net GHG emissions from conventional steam methane reforming based methanol is about 497 kilograms of CO_2e/tonne of methanol whereas the GHG emissions from the proposed bi-reforming based process is about 203 kilograms of CO_2e/tonne of methanol produced on an LHV basis. The proposed technology can reduce about a million tonnes of CO_2 equivalent GHGs per year in a 10,000 tonne methanol/day plant.

480271 Selecting Right Flare System to Meet Regulatory Requirments

Abstract:

Flare is considered as last line of defense in any process plant. This is one equipment which must be available at all time when the plant is in operation including during emergency cases. The purpose of the flare system is to reliably and safely dispose any and all hydrocarbon releases from the plant during startup, normal operation, maintenance and emergencies. The Federal Regulation (40 CFR 60.8) requires that emissions, except for emergency cases, shall be smokeless. 'Smokeless' flare operation is defined as generating emissions within 20% opacity levels, also known as Ringelmann#1 (R1) on Ringelmann smoke Chart. Typically, this smokeless flow tends to fall within 10 to 15% of the emergency flaring load for an LNG (Liquefied Natural Gas) plant. The apparent darkness or opacity of a stack plume depends upon:

- Concentration and size of the particulate matter in the effluent
- Depth of the smoke column being viewed
- Natural lighting conditions such as the direction of the sun relative to the observer, and Color of the particles

There are several limitations to select the right flare system. In the case study presented in this paper, four major criteria have been considered for evaluation. (a) total plot space availability / usage, (b) Flaring noise during normal operation, start-up & maintenance cases (c) Equipment cost and (d) Ease of maintenance. Smokeless characteristic depends on many factors including turbulence (characterized by exit velocity and mixing), molecular weight, heating value and chemical structure of waste gases and time available for combustion.

In the case study presented in paper for an LNG plant, the following type of flares are evaluated to optimize the flare selection, design, and cost.

- Elevated pipe flare including air / steam assisted flaring.
- Multi-point Ground flare (MPGF).
- Totally Enclosed Ground Flare or Enclosed Ground Flare (TEGF or EGF).

The project team needs to get a very good estimate of the amount of plot required for the flare system even in the Front End Engineering phase of the project. A detailed flare study is required to determine the type and cost of the flaring system.

This paper discusses various flare system designs available to achieve smokeless flaring during normal operation, start-up, and maintenance. Selection of right flare or right combinations of flare types are very important and must be evaluated carefully for overall plant reliability and cost effectiveness.

480334 Optimization of Shutdown Procedures of Olefin Plants

Abstract:

Shutdown and startup procedures are important operations for maintenance of olefin plants. Both procedures involve complicated facility operations, take long time and may produce huge amount of flare emissions. Optimize shutdown procedures could shorten duration time, reduce flare emissions and decrease cost by saving intermediate materials.

In this presentation, a new shutdown methodology for olefin plants is proposed. The key issue is to recover remaining liquid and vapor materials in facilities during the shutdown procedure. To achieve that, CGC (Charge Gas Compressors) must be kept running during recovering, and recovered hydrocarbon is sent to other plants as feed. Dynamic simulations of the whole procedure are used to determine possibility and to provide necessary operation data.

Moreover, the shutdown procedure can be considered as a scheduling problem. Thus, an MIP (Mixed-Integer Programming) model is developed to get an optimized shutdown schedule for facilities. By optimizing shutdown arrangements for facilities, the overall shutdown time could be reduced by 25%. Combining with the new shutdown methodology, over 90% of hydrocarbon could be saved, and flare emissions could also be reduced by 90%.

AUTHOR INDEX